THE FIELD GUIDE TO
Water wells and boreholes

The Geological Society of London Handbook Series
published in association with
the Open University Press comprises

Barnes: *Basic Geological Mapping*
Tucker: *The Field Description of Sedimentary Rocks*
Fry: *The Field Description of Metamorphic Rocks*
Thorpe and Brown: *The Field Description of Igneous Rocks*
McClay: *The Mapping of Geological Structures*
Milsom: *Field Guide to Geophysics* (in preparation)
Goldring: *Fossils in the Field* (in preparation)

Professional Handbooks in the Series

Brassington: *Field Hydrogeology*
Clark: *The Field Guide to Water wells and Boreholes*

Geological Society of London
Professional Handbook

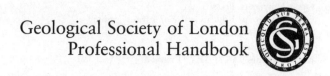

HANDBOOK SERIES EDITOR — M.H. de FREITAS

THE FIELD GUIDE TO

Water wells and boreholes

Lewis Clark

OPEN UNIVERSITY PRESS
Milton Keynes
and
HALSTED PRESS
John Wiley & Sons
New York-Toronto

Open University Press
Open University Educational Enterprises Limited
12 Cofferidge Close
Stony Stratford
Milton Keynes MK11 1BY, England

First Published 1988

British Library Cataloguing in Publication Data
Clark, Lewis
 The field guide to water wells and boreholes.
 —(Hydrogeology field handbooks).
 1. Wells—Design and construction
 I. Title II. Series
 628.1'14   TD405

ISBN 0-335-15203-1

Published in the U.S.A., Canada and Latin America by
Halsted Press, a Division of John Wiley & Sons, Inc., New York

Library of Congress Cataloguing in Publication Data

Clark, Lewis J. (Lewis John), 1937–
 The field guide to waterwells and boreholes.
 Bibliography: p.
 1. Wells—Design and construction. 2. Boring.
 I. Title.
 TD405.C52 1988   628.1' 14   87–35030

ISBN 0-470-21074-5 (Halsted Press)

Project management: Clarke Williams

Printed in Great Britain

Contents

The field guide to water wells and boreholes

Contents

Preface

This book is a practical guide to the work involved in drilling boreholes and wells. It should be useful in the office in the design stages of a project, but it is primarily intended to be of use to fieldworkers implementing the projects. The book is not a driller's manual; it is intended for people who design and supervise the drilling activities.

The correct design of boreholes and wells in relation to their hydrogeological setting is considered fundamental to any groundwater investigation and therefore well design and construction are dealt with in some detail. The emphasis in groundwater projects has changed to some extent from resources to pollution studies, and so attention is also given to the sampling techniques used with drilling. Pumping-test analysis is covered in detail in other texts, so that only the basis of pump testing of wells is considered here. A full reference list gives the reader guidance to those texts useful in the field, for design specifications and for general background reading.

Lewis Clark

Acknowledgements

Drs Arad and Faillace aroused my interest in hydrogeology while working with them in Karamoja, Uganda. My former colleagues at Groundwater Development Consultants Ltd, particularly Wictor Bakiewicz, began my education in the technology of boreholes and wells, and this has continued with my present colleagues at WRc Environment, Medmenham, Bucks, UK.

Mike de Freitas and Mike Packman read the draft of this book and I thank them for their helpful comments. Their suggestions have been used to improve the clarity of the text but any opinions expressed in the book are my own.

I owe much to the patience of my wife Joan while she corrected my English and to Nicholas who was very tolerant while training his father to use a work processor.

1

Introduction

This manual is intended to be a practical guide to the principles involved in the design and construction of boreholes and wells, and a source book of information useful to people planning groundwater investigations and supervising drilling in the field.

1.1 Terminology

1.1.1 Main types of wells and boreholes

Terminology is not standard throughout the world, and different names are commonly applied to identical constructions. In this Guide the following names are used:

1. *Water well* A borehole drilled for the principal purpose of obtaining a water supply. Synonyms are tubewell or production well/borehole.
2. *Hand-dug well* A large-diameter, shallow water well constructed by manual labour. Synonyms are open well or dug well.
3. *Test well.* A borehole drilled to test an aquifer by means of pumping tests.
4. *An aquifer* is a water-bearing geological formation capable of yielding groundwater in useful amounts.
5. *An aquiclude* is a geological formation with extremely low permeability.
6. *An aquitard* is a geological formation of low permeability but through which measurable leakage of groundwater can occur. In a system of aquifers separated by aquitards or aquicludes, each aquifer may have a different piezometric or hydraulic head and may contain water of a different quality.
7. *Exploration borehole.* A borehole drilled for the specific purpose of obtaining information about subsurface geological formations or the groundwater. Synonyms are investigation borehole or pilot borehole.
8. *Observation borehole.* A borehole constructed to obtain long-term information on variations in the groundwater level or quality.
9. *A piezometer* is a small-diameter borehole or tube specially constructed for the measurement of hydraulic head at a specific depth within an aquifer system or thick aquifer. The screened section in a piezometer is very short compared

with that in a normal observation borehole.

1.1.2 Groundwater flow

Groundwater under natural conditions flows from areas of recharge, normally the aquifers's outcrop area, to points of discharge at springs, rivers or in the sea (Walton, 1970; Freeze & Cherry, 1979; Todd, 1980). The driving force of this groundwater flow is the *hydraulic head:* the difference in level of the groundwater surface between the recharge and discharge areas.

The flow of water through the saturated zone of an aquifer can be represented by the Darcy equation:

$$Q = KiA$$

where Q = groundwater discharge rate (m³/day); K = hydraulic conductivity (m/d); i = piezometric or hydraulic gradient in the direction of flow; A = cross-sectional area through which flow takes place (m²).

The aquifer characteristics, which control the flow of water through an aquifer, are defined as follows.

Aquifer storativity   has two facets—unconfined storage or specific yield, and confined storage or the storage coefficient. The specific yield of an aquifer is the volume of water which will drain from a unit volume of aquifer under gravity alone. The storage coefficient of a confined aquifer is the volume of water released from storage per unit surface area of aquifer per unit change in head. This storage is related to the compressibility of water and of the aquifer fabric. The specific yield is usually about three orders of magnitude greater than the storage coefficient of the same aquifer.

The porosity   of an aquifer is the proportion of the aquifer material made up of voids. *Primary porosity* represents those voids present from the period when the sediments making up the aquifer were laid down, for example, the intergranular pores in a sandstone. *Secondary porosity* has resulted from changes in an aquifer after its formation, for example, fissures along joints or bedding planes. An aquifer in which groundwater can flow through the voids of both primary and secondary porosity is called a *dual-porosity aquifer.*

Hydraulic conductivity   is the rate at which *water* is transmitted through a unit cross-sectional area of aquifer under unit hydraulic gradient. The hydraulic conductivity is also commonly called the *aquifer permeability* or *coefficient of permeability.* Permeability so defined must be differentiated from the *intrinsic permeability* of a porous medium. The intrinsic permeability of an aquifer is independent of the liquid involved, is characteristic of the aquifer alone and is related to the hydraulic conductivity by the equation:

$$k = Kv/g$$

where k = intrinsic permeability (length²); K = hydraulic conductivity (length/ time);

v = fluid kinematic viscosity (length^2/time); g = gravitational acceleration (length/time^2)

The hydraulic conductivity of an intergranular aquifer will depend on the grain size and sorting of the aquifer material and the degree of cementation. In fissured aquifers the intensity of fissuring and the openness and continuity of individual fissures will control the hydraulic conductivity.

Transmissivity is the rate at which water can pass through the thickness of saturated aquifer of unit width under a unit hydraulic gradient. The transmissivity in a uniform aquifer will be the hydraulic conductivity multiplied by the saturated aquifer thickness but, as uniform aquifers rarely occur in nature, the transmissivity should be considered as a summation of the permeabilities of unit thicknesses through the aquifer section.

The natural flow conditions in an aquifer are disturbed when a well is drilled into the formation. The action of pumping water from the well lowers the level of groundwater and creates a hydraulic head difference between the water in the well and that in the aquifer. This head difference causes water to flow into the well and so lowers the hydraulic head in the aquifer around the well. The effects of pumping spread radially through the aquifer and can be most easily demonstrated by the lowering of either the piezometric surface or the water table as a *cone of depression* around the well. This cone can be seen and measured in observation boreholes drilled around the abstraction well.

A *pumping test* is the pumping of a well under controlled conditions so that the reponse of the abstraction well and the growth of the cone of depression can be measured. The analysis of the pumping test then can be used to evaluate the aquifer characteristics.

1.2 Boreholes and wells in groundwater investigations

Boreholes can represent a large capital investment in a groundwater investigation, therefore each water well or borehole should be drilled to an optimum design and at the location where it can best fulfil its function. Optimum borehole design depends on a detailed knowledge of the geological formations and subsurface geometry in the area to be drilled. This information is not always available, particularly in developing countries or in rural areas of developed countries, and a groundwater investigation usually has to be a phased operation designed to build up the necessary information (Brassington, 1988). This phased approach (described below) applies to both groundwater resource studies and to pollution investigations. Such a study will usually have the following phases, each successive phase building on the results of the previous ones:

1. Desk studies.
2. Exploration.
3. Detailed investigation.
4. Implementation.

3

1.2.1 Desk studies

A great amount of information on the geology and hydrogeology of developed countries is already available, and the collection and analysis of existing data can obviate the need for the exploration phase. In Britain the main sources of hydrogeological information include:

1. British Geological Survey (BGS): Published and unpublished geological and hydrogeological maps and reports. The National Well Records, also held by the Water Research Centre (WRC).
2. Regional water authorities: Comprehensive local well records, geological and hydrogeological maps and reports, aquifer protection maps.
3. Scientific journals: Hydrogeological papers and reports.

Hydrogeological data in most developing countries are sparser and more difficult to obtain. An energetic hydrogeologist, however, can find a surprising amount of data. Sources to search include:

1. National Geological Survey: Geological maps and reports.
2. National ministries, state organizations and development corporations: Well records and natural resource surveys. Reports of aid projects by the United Nations and the World Bank.
3. Scientific journals.
4. Consultants: Reports on water resource projects.

Consultant's reports can be the most valuable source of information, but they are often hidden on office shelves.

The information derived from all available sources should be tabulated in an orderly manner and plotted as a series of base maps of the area, which include:

1. Topographic and geological maps on the largest scale available.
2. Hydrogeological maps showing aquifer distribution, borehole locations and hydrographic features.

From these data, hydrogeological cross-sections should be prepared to show the subsurface geometry of the geological/hydrogeological formations.

1.2.2 Exploration drilling

The desk study will have shown the spread of available data and identified gaps in those data. A groundwater resource survey in a developing country will usually include an exploration phase to fill in the data gaps or, in some cases, to provide preliminary data on the area. Exploration drilling is rarely planned in isolation; it is usually just one aspect of an integrated project (Table 1.1). Aerial photographs or satellite imagery will be used to supplement existing geological knowledge, and geophysical surveys may be used to supplement existing data if necessary.

Exploration drilling is used to confirm the provisional interpretation of

Table 1.1 Flow chart for a ground-water resource investigation

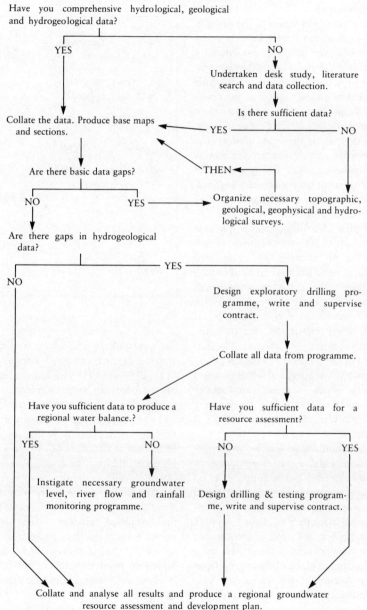

Have you comprehensive hydrological, geological and hydrogeological data?

YES

NO

Undertaken desk study, literature search and data collection.

Is there sufficient data?

Collate the data. Produce base maps and sections.

YES

NO

Are there basic data gaps?

THEN

Organize necessary topographic, geological, geophysical and hydrological surveys.

NO

YES

Are there gaps in hydrogeological data?

YES

NO

Design exploratory drilling programme, write and supervise contract.

Collate all data from programme.

Have you sufficient data to produce a regional water balance.?

Have you sufficient data for a resource assessment?

YES

NO

NO

YES

Instigate necessary groundwater level, river flow and rainfall monitoring programme.

Design drilling & testing programme, write and supervise contract.

Collate and analyse all results and produce a regional groundwater resource assessment and development plan.

The field guide to water wells and boreholes

the hydrogeological situation derived from the desk studies and the supporting exploration surveys. The siting of the exploration boreholes in a situation where there are a minimum of data is difficult, but should be governed by the principle that every borehole should be drilled to provide an answer to a question—for example: How thick is the aquifer? How deep is the aquifer? or, Is there an aquifer? The temptation to drill large numbers of boreholes should be resisted, for money spent on unnecessary exploration could deplete the budget of a later phase of the project.

Figure 1.1 illustrates a possible situation where a water supply is needed for an abattoir and associated holding pens. Little is known about the geology except that a sandstone dips to the south at about ten degrees beneath a clay formation. A river with a gravel floodplain crosses the area from the north-east to the south-west. The outcrop of the sandstone, which is clearly a potential aquifer, is about 200 metres wide, which suggests an aquifer thickness of about 34·7 metres or a drilled vertical thickness of 35·3 metres.

Assuming that this is the only information available, then five exploratory boreholes (BH 1–5) should be drilled at the locations indicated; these would provide the basic hydrogeological information for the area. Boreholes BH1 and BH2 in the river gravels would give the thickness of the gravels, provide samples of their lithology and give an indication of their variability. This information should be sufficient to show whether the gravels are a viable

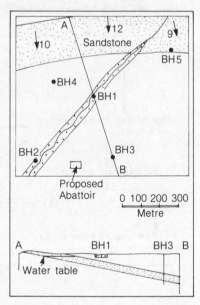

Fig. 1.1 Exploration drilling programme.

shallow aquifer and give data on which water wells can be designed. Boreholes BH3, BH4 and BH5 give information on the deep sandstone aquifer, particularly on its lithological variability and the groundwater elevations and flow direction. Borehole BH5 gives information on the unconfined zone and borehole BH3 on the confined zone. Borehole BH3 is important to verify the dip of the sandstone at depth and to show the nature of the aquifer close to the proposed abattoir. The dips measured at the surface range from 9 to 12 degrees, a range which, one kilometre from outcrop at borehole BH3 means a range in possible depths to aquifer of 160 to 210 metres. The lithological log of borehole BH3 will

6

give the actual depth to the aquifer and provide data for the design of future test wells and observation boreholes.

1.2.3 *Detailed investigations*

The detailed investigation phase is intended to supplement the exploration phase of an investigation by infilling data gaps and by quantifying the behaviour of the hydrogeological units. As with the exploration phase, drilling will be accompanied by other surveys such as surface geophysics.

It is important in resource studies to establish the relationships between the groundwater, surface recharge and surface water bodies. Observation boreholes are needed to follow the response of the groundwater levels to recharge events through the annual hydrological cycle. Observation boreholes adjacent to streams also can be used to observe the interaction between ground and surface water. Continuous observations of water levels by means of an automatic recorder are needed in at least one borehole. This record would then act as a base against which to compare water levels in other boreholes measured on a periodic basis—weekly or monthly.

The observation boreholes also should be used to obtain water samples for analysis to measure the variation in regional groundwater quality.

The values of the aquifer characteristics are needed to predict the effect of a particular pumping policy on the groundwater regime, and are obtained by means of pumping tests on test wells. Test wells, with their satellite

observation boreholes, are expensive installations and their number in any particular investigation is likely to be limited by budget constraints. In a uniform aquifer, test wells can be widely spaced—several or many kilometres apart, but in an aquifer which is variable, or in a situation where there are several interfingering aquifers, test wells may have to be close together before the results of their pumping tests can be used to predict aquifer behaviour with any confidence.

The quality of water from the test wells should be monitored on site during the tests, to detect any changes with time. At least one sample also should be taken for analysis for all major ions. This sample, taken towards the end of the test, will show the quality of the water and its suitability for future uses.

It should be remembered that a pumping test, even a long test lasting several weeks, is probably testing only a few square kilometres of aquifer and that the knowledge of the aquifer characteristics obtained from such a test is by no means perfect. Also, aquifer behaviour over several years may have to be predicted from observations taken over a period of only a few weeks.

The minimum number of test wells needed is one for each significant facies change in the aquifer being studied, and one for any specific interaction which needs to be observed. The area shown in Fig. 1.1 can illustrate the extension of the exploratory phase through the detailed investigations. In this area, geophysical surveys indicated that the eastern part of the sandstone aquifer is much finer than the western part (Fig. 1.2). Drilling results from exploration hole BH 5 and a new observation borehole BH 6 confirm this facies change. Observation boreholes BH 7 and BH 8 provide further lithogical data and permanent water-level/quality monitoring points in the unconfined and confined parts of the target aquifer.

Fig. 1.2 Detailed drilling programme.

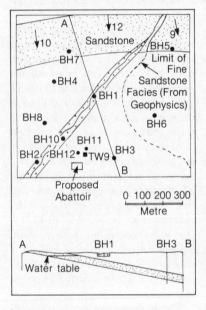

In this case there is a single demand point—the abattoir—so, initially, only one test well would be proposed—TW 9, with two satellite observation borehole. One additional observation borehole would be drilled in the alluvium to measure the interaction, if any, between the deep aquifer and the alluvium during the pumping test on

TW 9. In a general resource study a further test well would be located near BH 6 to test the finer part of the aquifer, and possibly one in the alluvium to test that formation.

1.2.4 Implementation phase

The planning of the implementation phase of a groundwater resource project will depend to some extent on the ratio of demand to the ability of a single source to supply that demand. The latter will have been measured by the pumping tests of the site investigation phase. In the case of a minimum demand, where one source could theoretically supply the demand, the production pumping station should be designed with a minimum of two water wells: one for production and one for stand by.

Where more than one borehole is needed to cope with demand, the well field should always include at least one stand-by borehole. The closer the water wells are spaced, the greater is their mutual interference, and therefore the greater are their running costs, but on the other hand, the costs of pipelines connecting the wells in the field are decreased. The distances between wells in the well field should be optimized by matching the water-well capital and running costs against pipelines costs (Section 2.1.4).

The treatment works for the water to be discharged from the production well field probably will have been designed before the well field comes on-line. The chemical data on which the works are designed will have been obtained by analyses of water from the pumping tests of the detailed investigation phase. It is important in that phase to ensure that the quality of water supplied, together with its designed treatment, will give a water of suitable quality for its proposed use. In the implementation phase the groundwater quality should be monitored to ensure that the quality criteria on which the treatment works were designed do not change with time.

2
Design of well and borehole structures

The summary of the drilling activities involved in a groundwater investigation (Section 1.2) illustrates the diversity of functions of boreholes. Each borehole or well must be designed to fulfil its specific function efficiently and at the least cost. The borehole design(s) have to be chosen before drilling begins, as they are the basis of a drilling contract. The designs, therefore, have to be based on existing information, and the more comprehensive this information is, the more successful will be the design. Investment in the early phases of an investigation — the desk study and the exploration phase — will always be repaid.

2.1 Water wells

The design of a water well depends on the type of aquifer system being exploited and the discharge rate required. The discharge rate (or supply demand) must be decided before a water well can be designed, because it will dictate the size of pump required which, in turn, will govern the

minimum internal diameter (ID) of the pump-chamber casing. The cardinal rule of well design is that the ID of the pump chamber must be large enough to accommodate the pump, pump shroud, cable and rising main.

Aquifers have been divided into three broad classes for the purpose of water-well designs.

1. Crystalline aquifers.
2. Consolidated aquifers.
3. Unconsolidated aquifers.

Crystalline aquifers are typified by the igneous and metamorphic rocks which underlie large areas of the world. They have no primary porosity or permeability and have no well-defined base. Water-bearing voids in them are usually in the form of fissures. In an open borehole these rocks usually do not need support.

Consolidated aquifers are sedimentary formations that have sufficient strength to stand in an open borehole without support. They retain some of their initial — or primary — porosity and permeability, and normally have clearly defined tops and bases (Fig. 1.1). Many regional sandstone and

Fig. 2.1 Bunter Sandstone aquifer outcrop.

Fig. 2.2 Lower Greensand aquifer; an unstable formation is collapsing.

Fig. 2.3 Old Red Sandstone in Scotland.

limestone formations, for example the Bunter Sandstone of Western Europe, are consolidated aquifers (Fig. 2.1).

Unconsolidated aquifers are sediments which cannot stand in an open borehole, so that any well in such a formation will require support by casing and screens. These aquifers include alluvium and many Tertiary or Mesozoic sandstones (Fig. 2.2).

There are exceptions or transitional cases in such a classification. Sedimentary strata may become so compacted that they lose their primary porosity and become entirely crystalline, for example quartzites or marbles. Indurated Palaeozoic sandstones such as the Old Red Sandstone have many features of crystalline formations (Fig. 2.3).

7 m section

Table 2.1 Classification of aquifers in the field.

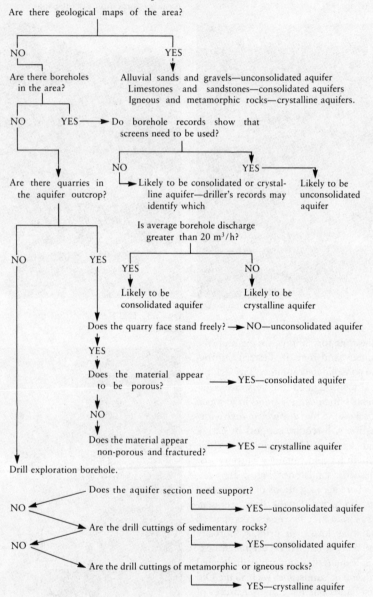

Finally: geophysically log the borehole to confirm the aquifer classification.

Basalt successions commonly behave like consolidated sedimentary aquifers rather than crystalline formations. Parts of consolidated aquifers may lose their cement through solution and revert to the character of unconsolidated aquifers.

The bases for the classification of aquifer types in the field are given in Table 2.1.

2.1.1 Wells in crystalline aquifers

Groundwater occurs in crystalline aquifers in the secondary porosity caused by weathering and fissuring. Since both weathering and fissuring decrease with depth there will be a depth beyond which the cost of drilling outweighs the chance of significantly increasing the yield of a borehole. This depth is the optimum total depth of a water well and will vary from place to place depending on the geological and geomorphological history of the site. It is unlikely, however, to exceed 100 m. (Clark, 1985). The optimum depth at any location has to be determined from previous drilling experience in the area, or from surface geophysical surveys which can indicate the bottom of the weathered zone.

The yield of water wells in crystalline aquifers is low, averaging about 50 m^3/d and rarely exceeding 250 m^3/d, so large diameters are not needed in such wells. A pump with 150 mm outside diameter (OD) will cope with the available discharge from almost all water wells in crystalline rocks, and so a 200-mm ID pump

chamber should be adequate in their construction. In most situations only the upper few metres need casing and grouting to seal off surface water, and the rest of the well can be left open (Fig. 2.4A). Care has to be taken when completing a water well in crystalline rocks, so as not to case-off the more prolific shallow water-bearing zones just because they are unstable. Where these water-bearing zones are shattered rocks or incoherent weathered rock then a screened section can be incorporated in the design (Fig. 2.4B). If the screen is merely to support shattered rock; then the screen slots may be very coarse, but where incoherent weathered rock is screened then the slot size has to match the grain size of this granular material (Section 3.2).

2.1.2 Wells in consolidated aquifers

Single aquifers. A water well designed to exploit an aquifer of limited thickness will be drilled to penetrate the whole aquifer. In an area where there has been drilling in the past, experience will indicate the potential discharge rate of a well, but in virgin territory, test wells will be needed to provide this information. The potential discharge rate and the amount of lift required (depth to static water level plus potential drawdown of the water level in the well), can be used to dictate the minimum ID of the pump-chamber casing, which should be at least 50 mm greater than the OD of the pump or rising main.

13

Fig. 2.4 Well designs for crystalline aquifers.

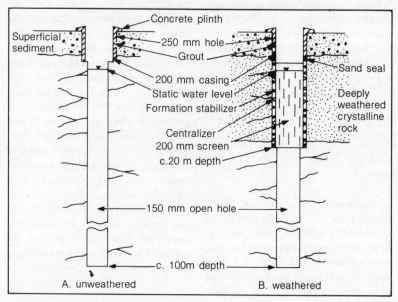

In shallow aquifers, water wells are usually drilled in two stages. A borehole is drilled at a diameter about 50 mm greater than the pump chamber OD to about 2 m into the target aquifer, and then the pump chamber is set and grouted into position (Fig. 2.5A). Drilling then continues to total depth for an open hole completion. Water wells in deep aquifers may be drilled in several stages, with intermediate casing strings between the pump chamber and the aquifer (Fig. 2.5B). The pump-chamber casing diameter chosen is then a compromise between the minimum needed to hold the pump and the diameter needed to allow the intermediate casing strings to pass through. The diameter of the in-termediate casing must be large enough to allow the drill bit to pass for the final drilling. The diameter of the final section through the aquifer has to be chosen with care, for it has direct effects not only on the potential discharge of the well but also on the diameters of all the casing strings, and therefore on the well costs. A minimum diameter of about 150 mm is recommended as this allows most work-over tools, such as jetting tools, in the hole for well maintenance (Chapter 10). The increase in diameter of drilling, however, does not give a proportional increase in yield of the well (Fig. 2.6) and large diameters are rarely justified on economic grounds.

The design of a water well in a very thick, consolidated aquifer is similar to

Fig. 2.5 Well designs for consolidated aquifers.

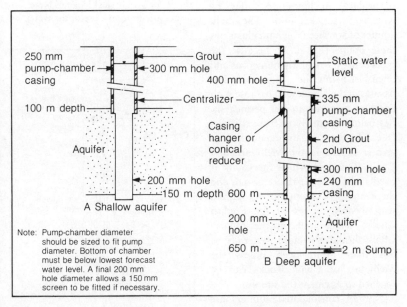

that in a crystalline aquifer, because there is no well-defined base to the aquifer. Also, the groundwater flow in almost all consolidated aquifers is largely through fissures or zones of enhanced permeability, so criteria for optimizing well design, based on a uniform aquifer (Section 2.1.3), cannot be applied. The depth of borehole needed for a specific discharge will depend on the distribution and size of water-bearing fissures — an aquifer feature that can rarely be predicted. Well design has to depend almost entirely on experience of previous drilling in the area, although some broad generalizations can be made. The fissures are mainly bedding features, and tend to be more frequent and more open near to the surface. In addition,

Fig. 2.6 The theoretical increase in yield with increase in well diameter. Expressed as % increase over yield of a 150 mm diameter well. After Driscoll, 1986, table 13.15.

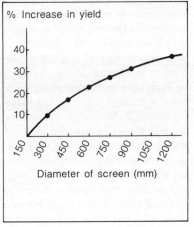

those fissures near to the water table, particularly in limestones, may be widened by active solution. The Chalk aquifer of Southern England illustrates these points (Fig. 2.7), because the fissures are particularly numerous and open just below the water table. The fissures in the Chalk tend to close with depth to give low aquifer permeabilities below about 80 m from the surface. The fissures can remain open at greater depths in the Melbourn and Chalk Rocks, in which the Chalk is harder and stronger than normal, and these are two commonly recognized water-bearing zones of the English Chalk. Water wells in the English Chalk should, therefore, generally be less than 80 m deep unless the Melbourn or Chalk Rocks can be reached to augment the supply.

Not all consolidated aquifers are completely stable, and this can give rise to problems. Another factor to be taken into account, where there is fissure flow within the aquifer, is erosion of sediment from the fissure walls by rapid flow along the fissures. This sediment in the water discharged from the well can damage the pump and silt-up surface works. The judgement on long-term formation stability is difficult to make, and can only be based on experience with other water wells, either in the area or in similar aquifers elsewhere. If there is any doubt, then the well should be designed to incorporate the necessary screens and gravel pack, just as if the aquifer was an unconsolidated sediment. The design specifications for screens should be finalized at the design stage, before a drilling contact goes to tender. It is

Fig. 2.7 Fissures in Chalk in Kent, UK. Shown by geophysical logging. Note their concentration close to the water table.

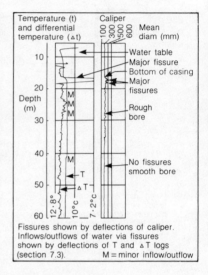

Fissures shown by deflections of caliper. Inflows/outflows of water via fissures shown by deflections of T and ∆T logs (section 7.3). M = minor inflow/outflow

both difficult and expensive to adapt a borehole not designed for screens, and to keep it open while the screen is designed, ordered and supplied.

The sequence of steps that a person will have to follow in designing a borehole or well is summarized in Table 2.2. This sequence will need to be followed before a drilling contract can be prepared adequately.

Multiple aquifers. A water well in a multiple aquifer system (Fig. 2.8) will almost certainly have to be completed as a screened well, because the intervening aquicludes (impermeable formations) in the system will tend to be unstable and need casing-off. In theory, if the aquifers really are consolidated then the screen design is irrelevant but, in practice, to avoid any

Table 2.2 A summary of steps to follow in the design of a borehole or well

---

Firstly:
1. Have you identified the aquifer and succession to be drilled?
2. Have you checked the records of existing wells, boreholes and investigations in the area?
3. Have you decided the purpose of the borehole or well?—exploration, observation or production well.

If so, you have the basis for designing the structure, and you then have to decide:
a) What depth and diameter will the borehole be?
b) Will it need to be cased in part or entirely?
c) Will the aquifer need screening?
d) Will a gravel pack be needed or will a formation stabilizer be sufficient?

When the structure has been established then the materials to be used in the construction should be chosen:
a) What material is needed for the casing?
b) What material is needed for the screen?
c) What material is needed for the gravel pack? The pack will need to be designed from information available on the aquifer.
d) What design of screen is needed? For example, wire-wound, bridge slot or machine slotted.
e) What screen slot size is needed?
f) Does the groundwater quality make special cements necessary for grouting? For example, sulphate-resistant cement.

Sampling is necessary to determine accurately the geological succession and the depth and character of the aquifer. The sampling features that it is necessary to define are:
1. Depth intervals of disturbed formation samples.
2. Method to be used to collect disturbed formation samples.
3. Depth intervals of undisturbed formation samples (cores).
4. Method of collecting undisturbed formation samples.

The factors listed above, together with local conditions, will be used to determine how the borehole or well is to be drilled. The next decisions are to:
a) Decide on the best method of drilling.
b) Decide on the drilling rig to be used.

---

possibility of sand-sized particles entering the well, the screens should be chosen on the assumption that the aquifer material is unconsolidated. The screen sections should end at least one metre from the top and bottom of each aquifer, to avoid incursion of aquiclude material. The annulus around the screen string can be infilled with gravel pack material — to act as a formation stabilizer to prevent undue collapse of material down the annulus, and to avoid turbulence behind the screen if the annulus is left open. The emplacement of a formation stabilizer may be difficult in deep boreholes.

Fig. 2.8 A well design for a consolidated multiple aquifer system.

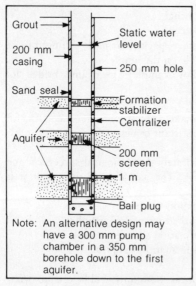

Grout
200 mm casing
Sand seal
Aquifer

Static water level
250 mm hole
Formation stabilizer
Centralizer
200 mm screen
1 m
Bail plug

Note: An alternative design may have a 300 mm pump chamber in a 350 mm borehole down to the first aquifer.

The accurate emplacement of screens in a multiple aquifer system is essential, and, in almost all cases, will need geophysical formation logging (Section 7.3) to provide the necessary accuracy in depth control.

2.1.3 Wells in unconsolidated aquifers

The most common unconsolidated aquifers are alluvial deposits along river floodplains or terraces, and these range from thin gravel beds along small rivers, to multi-aquifer systems several hundred metres thick — along such major rivers as the Indus. The design of water wells in unconsolidated aquifers has much in common with that of wells in consolidated strata, except that the former invariably require screening to prevent formation collapse. A water-well design varies with the number of aquifers to be exploited and the depth of those aquifers, but each design will incorporate the following features: total depth, drilled diameter, casing selection, screen selection, casing/screen installation and gravel pack design (Fig. 2.9).

Total depth. In an aquifer of limited thickness a well would be fully penetrating so as to maximize the yield.

The depth of a well in a very thick aquifer is governed by cost constraints or by the discharge demand. The thickness of aquifer to be drilled to give a specific discharge can be calculated approximately using Logan's steady-state equation (Kruseman & De Ridder, p. 170), provided that the hydraulic conductivity is uniform through the aquifer thickness:

Fig. 2.9 A well design for an unconsolidated aquifer.

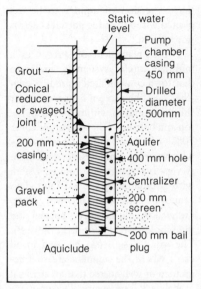

$$kD = 1 \cdot 22Q/Sw$$

where k = hydraulic conductivity, m/day; D = saturated thickness of aquifer to be drilled (m); Q = proposed discharge rate (m³/d); Sw = drawdown (m); $1 \cdot 22$ = constant, assuming no well losses; $2 \cdot 0$ can be used for design estimates.

As an example, with a well where a discharge of 2500 m³/d is needed, where a drawdown of 25 m is acceptable and the hydraulic conductivity is 5 m/d, then

$$D = 2 \cdot 0 \times 2500/25 \times 5 = 40 \text{ m}$$

that is, a minimum penetration of 40 m of saturated aquifer is likely to be needed in the well.

Drilled diameter. The drilled diameter is dictated by the casing design chosen for the well. The drilled hole at any depth should have a minimum diameter about 50 mm greater than the OD of the casing/screen string. The diameter of a borehole to be equipped with an artificial sand or gravel pack has to be about 200 mm greater than the screen OD to accommodate the pack.

Casing/screen size and location. The casing in shallow wells is in one string, with an ID capable of accommodating the pump. The screen may be the same diameter as the casing or smaller (Fig. 2.9). The latter design can result in savings in capital costs.

In deep wells the casing may be in two or more strings — the pump chamber and the intermediate strings as in Fig. 2.5B. If the various casing/screen sections are to be installed in separate operations, then it is essential to ensure that they will nest inside each other. The intermediate casing must be large enough for the screen to pass, and be small enough to pass through the pump-chamber casing. The pump chamber has to be large enough to accommodate any proposed pumping equipment.

The bottom of the pump chamber commonly defines the depth below which the pump cannot be lowered, therefore it must be at a safe depth below the static water level in the well. This distance is the drawdown expected after several months of pumping at the production rate, plus the height of the pump and a safety factor. The latter is based on a judgement on the long-term behaviour of the water

19

The field guide to water wells and boreholes

table in the area being exploited, over the lifetime of the borehole. This safety factor can be greater than the other two factors put together. In many areas of the world, a feature of sedimentary basins is the abandonment of wells as their pump chambers are dewatered. A safety factor of at least double the maximum expected drawdown is recommended.

The decisions to be made on screen selection involve the length, diameter and type of screen to be used. The screen is usually more expensive than the casing, so the efficient use of the screen can minimize the capital cost of a well. The screen should not extend above the drawn down level while the well is in production. This avoids the upper section of the aquifer and screen becoming aerated as the aquifer is dewatered, and reduces the risk of incrustation of the screen and aquifer by iron from the groundwater.

In the case of a single aquifer of limited thickness, it is generally recognized that a screen covering at least 70% of the aquifer thickness will perform almost as effectively as if the whole aquifer was screened (Johnson 1966, p. 184). This allows for casing to be set into the top of the aquifer and a bail plug to be put at the bottom of the screen. The bail plug is a short length of casing, capped at the bottom to allow the well to be cleaned out without damaging the screen. A screen covering much less than 70% of the aquifer thickness should be avoided, as partial penetration effects will decrease the yield of the well.

The total screening of a thick aquifer may not be feasible, but a spacing of short lengths of screen, separated by lengths of blind casing, enables a thicker section of aquifer to be exploited without undue partial penetration effects.

The emplacement of screens in a multiple aquifer system is similar to that for a consolidated aquifer system (Fig. 2.8). The screen for *each* aquifer should extend to within 1 m from the top and bottom of the aquifer, or cover 70% of the aquifer thickness — whichever is the greatest screen length. A very heterogeneous aquifer system with alternating thin sands and clays cannot be screened in this way. Here, the well should be gravel packed and the whole system screened (Fig. 2.9). The same design can be used in all cases where the aquifers are so fine-grained or well-sorted that an artificial gravel pack is necessary. The grain size of the gravel pack (Chapter 3) should be selected to suit the finer-grained sand beds, in order to minimize the danger of drawing sediment into the well.

The diameter of the screen should be at least 150 mm, to allow ready access for work-over tools in future maintenance operations. Screen diameters in excess of 300 mm, however, are rarely justified, particularly in deep wells.

Screens are manufactured to many designs in a wide range of materials. The question of choice of these screen *types* is dealt with in some detail in Section 3.3.

Gravel pack choice. A gravel or sand pack is introduced around the screen of a water well, to produce an envelope of material with enhanced permeability and physical stability adjacent to the screen. The enhanced permeability reduces well losses (p.99) and incrustation of the screen (p.104), while the physical stability reduces the amount of sediment drawn into the well by pumping. Two main kinds of pack are used, natural and artificial, depending on the kind of aquifer being drilled.

A natural gravel pack is produced by the development of the aquifer formation itself (Section 8.3). An aquifer is suitable for the development of a natural gravel pack if it is coarse grained and poorly sorted. Artificial gravel packs (Fig. 2.9) are used in unconsolidated aquifers, where the aquifer material is either very fine or well-sorted. A great advantage of the artificial pack is that, because the material is coarser than the formation, screens with larger slot sizes can be used. Artificial packs are useful in allowing laminated heterogeneous aquifers to be screened much more safely than with direct screening and natural development. The different criteria used for the development of natural gravel packs and the design of artificial gravel pack material are discussed in Section 3.2.

2.1.4 Economics of well design

The economic optimization of well design in most situations is dependent on common-sense guidelines to avoid over-design, because the total depth of a well and the length of screened sections are dictated by the aquifer geometry. These guidelines include the following:

1. Do not drill deeper than necessary.
2. Do not drill at a larger diameter than necessary.
   a) Do not design a gravel pack thicker than needed.
   b) Do not design for a screen or casing diameter greater than necessary.
3. Do not use expensive materials where cheaper ones will do.
4. Do not use more screen than is necessary.

As examples of these guidelines: in crystalline rocks a maximum depth for a well is about 100 m (Clark, 1985). Unless there is a known water-bearing structure at depth, drilling two shallow wells makes better economic sense than one deep well, as the chances of hitting water are greater.

A household demand is unlikely to exceed 1 m^3/day (220 gal/day), so it is a waste of money to design a high-yield well for such a house. A household can be supplied through a 100 mm OD pump. The installation of a pump chamber with an ID more than 200 mm would be unwarranted.

The well design in a thick, uniform, unconsolidated aquifer is more open to optimization techniques, because the aquifer geometry dictates neither the total depth nor the screen length in the well. The only design-aim to be met is the required discharge rate or demand at the lowest cost. The cost of the water pumped from a well depends upon both the capital costs and running costs of the well (Stoner *et al.*, 1979). These costs are interdependent because the components of a well design can affect the running (pumping) costs. An increase in screen length, for example, will increase the capital costs but — because of decreased drawdown — will decrease the running (pumping) costs. Similarly, a decrease in well diameter will decrease capital costs but may increase running costs through increased well losses. There is an economic optimum (i.e. least cost solution) for each design feature in a well.

The cost of a water well can be expressed as functions of various features in the well, for example, screen length, well depth, casing length and pump-chamber length (Fig. 2.9), and these costs can be reduced to 'present values' by normal discounting methods. The well performance is represented by similar functions of the well design and used to evaluate Logan's equation. (p.16). The present values of capital and recurring costs are combined, and, for each parameter to be examined, partially differentiated and equated to zero to determine the value of that parameter for minimum cost. An example of the derivation for optimum screen length of a well is shown on Fig. 2.10.

This approach to water-well design by Stoner *et al.* is an important guide to economic optimizations, but it does presuppose a uniformity of aquifer conditions and a control over economic factors which may not be present in that part of the world in which the hydrogeologist working.

Fig. 2.10 Example of the optimization of a screen length in a water well. After Clark & Stoner, 1979.

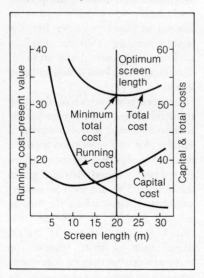

2.2 Hand-dug wells

The hand-dug well in Great Britain now has the image of a romantic wishing well in the back garden. Nevertheless, it must be realized that in rural areas such wells were the sole source of water for many families before the post-war spread of mains supplies. In developing countries the hand-dug well is still the most common technique of groundwater exploitation, and, as a source of potable water, it is still probably more important than drilled wells, and certainly more healthy than surface sources. The implementation of good well-design criteria therefore can have global effects on water supplies and public health. Excellent handbooks on hand-dug well construction are available, including those by Watt & Wood (1977) and DHV Consulting Engineers (1979).

2.2.1 Design for yield

Hand-dug wells have to possess a large enough diameter to accommodate the person making the excavation, but should not be wider than necessary. Excessive diameters increase the volume of debris to be removed, the time needed to dig the well, and the chance of surface pollution; all without significantly increasing the yield of the well.

The lining of a hand-dug well is made most commonly of permeable concrete rings or of masonry, but can include large-diameter metal or wooden casing and screen. In stable rocks the well is commonly left unlined. The method of installing linings varies, but two examples will give an indication of the common methods. In soft formations where the water table is relatively shallow, the aquifer is lined with porous concrete rings, while above the water table the well is lined with impermeable rings. The rings are about $1 \cdot 25 - 1 \cdot 5$ m in diameter, 1 m high and capable of being handled by a small crane. The bottom ring may have a cutting edge attached. A hole is dug from the surface to accommodate the bottom ring which is positioned carefully in a vertical position. Excavation continues from beneath the liner which follows the excavation under its own weight. Extra rings are added at

the surface as necessary. Excavation continues as far as possible below the water table, by lowering the water level in the well by pumping during the excavation. On completion, the joints in the rings are cemented and the surface works are built. The bottom of the well is stabilized with a layer of about 200 mm of coarse gravel (Fig. 2.11A). The well may be deepened further during the dry season when the water level is naturally at its lowest. In this event, the extension of the well will have to be lined by slimmer porous rings which can be telescoped inside the first well.

Where the water table is relatively deep and the ground is competent (Fig. 2.11B) the shaft to the water table can be excavated, then lined by masonry or, if the ground stability is uncertain, the masonry can be added course by course to the top or bottom of the lining as excavation progresses. On reaching the water table, the lining is pointed to give it strength and water tightness. The annulus behind the lining is filled with cement grout to stop surface water percolation. Excavation then continues as far as possible, using permeable concrete liners as in the first design.

In very marginal aquifers the hand-dug well is invaluable, for it acts as a reservoir into which the aquifer can leak continuously in order to fulfil a peak demand much greater than the instantaneous aquifer yield. The well is deepened to create a reservoir volume capable of either meeting demand or holding the daily yield of the aquifer.

23

The field guide to water wells and boreholes

Fig. 2.11 Hand-dug well completions.

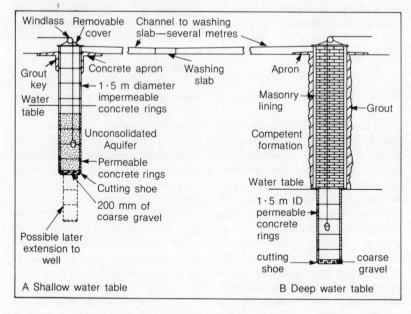

A Shallow water table

B Deep water table

2.2.2 Design for health

A hand-dug well is far more vulnerable to pollution than is a drilled borehole. The dug wells tend to be shallow and open to infiltration of polluted surface water, their lining and surface works are commonly badly finished so that spilled water or animal wastes at watering wells can flow back into the wells, and usually their tops are open, allowing rubbish to fall in.

A hand-dug well can be protected from these hazards by using the following design criteria to limit pollution of groundwater (Fig. 2.11):

1. Locate the well up the groundwater gradient (usually uphill) from any cess-pits or privies. Dig it at least 30 m from, or as far as possible from, the pollution points.

2. The upper lining of a well must be impermeable and should cover at least the upper two metres of the well.

3. The surface works of a well must shed spilled water away from the well, e.g. by a surrounding concrete apron sloping away from the well.

4. The surface works must encourage water to be used away from the well. Provide wash-basins several metres from the well.

5. The surface works must prevent effluents from entering the well. A wall must surround the well and the apron must be securely keyed to the well lining.

6. The surface works must prevent rubbish from falling into the well. A lid is needed or a buried completion. (DHV Consulting Engineers, 1979, p. 60).
7. The method used to lift water should discourage surface spillage. Several designs of windlass are available—a simple roller in common use in West Africa was designed to squeeze the water from the hoist rope and prevent transmission of guinea worm—a parasite occurring in some wells in the area—to the hands of the haulers. See Fig. 2.12.
8. The well should be disinfected on completion and then annually.

The well should be pumped to evacuate the bulk of the contaminated water and then filled with chlorinated water. Sufficient chlorine should be added in the form of hypochlorite to ensure that the water has a significant surplus of chlorine. The chlorine concentration should be above 1000 mg/l free Cl^2, but in those situations where the chlorine cannot be measured, after stirring with buckets or a recycling pump, the water should smell strongly of chlorine. The well should be left overnight if possible and then pumped to waste until the taste of chlorine has disappeared.

2.3 Rannay type wells

The Rannay well design can be considered as a variation of the hand-dug well, but usually on a larger scale. A large chamber or well is sunk into the aquifer, and horizontal boreholes are drilled radially outwards from this central well. Casing and screen are then jacked into the boreholes to form horizontal water wells. The design has the effect of increasing enormously the effective diameter of the well and therefore its yield.

Rannay wells are most commonly used in thin alluvial aquifers, where the central well can be several metres across and the horizontal boreholes can be several tens of metres in length. The central well is completed like a hand-dug well, and can be lined with masonry or concrete. The design of the horizontal boreholes follows the same criteria as a vertical water well in similar aquifer material. The Rannay design is useful for sites adjacent to rivers, because the horizontal boreholes can be directed beneath the river to induced recharge from the river—thereby increasing the yield. This feature of a Rannay well can be used to reduce water treatment costs by filtering the induce recharge through a natural fine-grained aquifer. A similar design can also be used in weathered or fissured rocks to increase the yield of open wells.

Fig. 2.12 Hand-dug well in Nigeria.

2.4 Observation boreholes

Observation boreholes tend to fall into three categories: those intended to provide a regional monitoring network; those drilled in a group around a test well for a pumping test; and those drilled for pollution investigations. In remote areas, exploration boreholes can be completed as observation boreholes, and, in the case of boreholes drilled for a pumping test, the first will be drilled as an exploration borehole to provide lithological data on which the other boreholes can be designed.

The design of an observation borehole will depend on the succession to be drilled, the nature of the aquifer under observation and the depth of borehole to be drilled. The design criteria in many cases will be simpler than those used for water wells in similar situations. As pump chambers are not required, observation wells can be completed to a much slimmer design than water wells. The completed internal diameter, however, must be large enough to accommodate any proposed sampling equipment or water-level recording equipment, and therefore 100–150 mm ID is usually recommended. The drilled diameter of the borehole is usually 50–100 mm greater than the maximum OD of the casing string.

The materials used in the construction of observation boreholes in developing countries may be of local origin: such as coir-wrapped bamboo screens, but the vast majority of observation boreholes will be made using similar materials to those in the water wells. Observation wells, up to about 200 m deep, can be completed using plastic casing and the expense of fibreglass or steel is justified in such wells only in exceptional cases. In pollution studies, for example, when organic contaminants are being investigated, plastics must be avoided because leachate from the plastic can interfere with the results. In tropical countries some plastics can deteriorate rapidly through solar degradation when stored in the sun, and could also suffer damage when transported over long distances on bad roads. In such situations, steel could be chosen because of its superior resistance to abuse.

An observation borehole is not designed to maximize its yield, so the screen need not extend the full thickness of the aquifer. The head within a single aquifer, however, can vary with depth and the screen distribution should take account of this. The hydraulic head shown by an observation borehole, screened for the full thickness of an aquifer, will be some average of the heads distributed through the aquifer. A similar average head will be given by a series of short screens distributed through the aquifer. In an observation well for a pumping test, it is important that the screen distribution should cover the same aquifer section as the screen in the production well.

A sedimentary aquifer system may comprise several aquifers separated by aquitards or aquicludes. This presents a problem in borehole design which has been resolved by either single or multiple completion.

Single completion. The observation borehole is completed with a single screen set adjacent to each aquifer (Figure 2.13A). The head and water quality observed are then some average value of those parameters in all the aquifers.

With an observation borehole for a pumping test in a multiple aquifer system, it is imperative that every aquifer screened by the test well is also screened in the observation borehole. If one or more aquifers are missed then a pumping test must overestimate the aquifer transmissivity, because the observation well will be responding to only a part of the discharge from the test well.

Multiple completion. The definition of the head distribution in an aquifer system, and therefore the understanding of the groundwater regime in the system, is improved in a multiple completion where a nest of piezometers is installed in a borehole. Each piezometer has a single screen adjacent to a single aquifer, and each screen is separated from the rest by an impermeable seal (Fig. 2.13B).

This method of completion, however, is more complex and much more expensive than a single completion. It may be impossible to satisfactorily instal the nest of piezometers in deep observation boreholes, because of the difficulty in setting the impermeable seals accurately. Installing a nest of tubes in a borehole also means either that the tubes have to be of a small diameter or the borehole has to be much larger than normal. In a pumping test the measurements from a multiple observation well are difficult to analyse, because of the need to determine the proportion of discharge derived from each aquifer. The result of these problems is that multiple completions tend to be restricted to relatively shallow aquifer systems.

An alternative to multiple completion in a single borehole is to drill a group of boreholes, each screened in a different aquifer. In this case the main problem is the cost of multiple drilling and, again, the technique tends to be restricted to shallow aquifer systems or specialist groundwater studies. Slim piezometers have been developed to study the head distribution in such relatively shallow aquifers. Many designs are available, but they all follow the same basic principles, a

Fig. 2.13 Observation borehole completions.

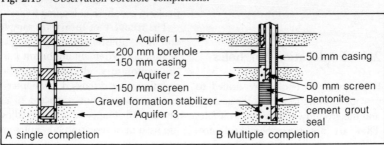

Aquifer 1
200 mm borehole
150 mm casing
Aquifer 2
150 mm screen
Gravel formation stabilizer
Aquifer 3
50 mm casing
50 mm screen
Bentonite–cement grout seal

A single completion   B Multiple completion

short screen tip is set at the end of a piezometer tube. The tube is usually a 12·5 or 25 mm ID pipe, either plastic or steel, while the screen can be a perforated section of pipe wrapped with filter gauze, or a purpose-made porous ceramic tip. The well-known 'Casagrande' type of piezometer tip is shown in Fig. 2.14.

Fig. 2.14 Casagrande-type piezometer tip.

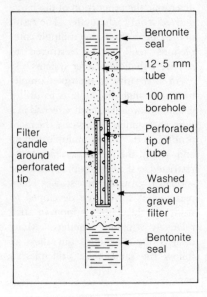

Bentonite seal

12·5 mm tube

100 mm borehole

Perforated tip of tube

Filter candle around perforated tip

Washed sand or gravel filter

Bentonite seal

2.5 Exploration boreholes

Exploration boreholes are drilled to provide information about the geology and the groundwater beneath the site. They are not permanent structures unless completed as observation wells,

and on completion are either backfilled or allowed to collapse. The correct completion of an exploratory borehole, particularly in a groundwater pollution investigation, is extremely important. A borehole left open can be a conduit for pollution from the surface to the groundwater body. Quite apart from polluting the groundwater, such conduits could give a false impression of the velocity of pollution movement and the degree of pollution attenuation, in a pollution study. It is recommended, therefore, that all exploration boreholes, when abandoned, are backfilled by pressure grouting through a tremie pipe set to the bottom of the hole.

The proposed depth of an exploration borehole may be specified in terms of a geological horizon, the bottom of the sandstone in Fig. 1.2, for example, but an estimate of the total depth must be made, as it affects the choice of drilling rig and possibly the method of drilling. The diameter of an exploratory borehole, because it may not be a permanent structure, is not critical. The drilled diameter will be determined largely by the depth of the hole and the type of samples needed. A minimum diameter of 150 mm will allow access for most sampling equipment, but particularly geophysical logging tools.

The recovery of formation samples and the establishment of a lithological section are extremely important parts of exploratory drilling. Sampling techniques are discussed in Chapter 5, but sampling is not restricted to exploratory drilling, and similar methods are used in observation boreholes and water wells.

3

Well and borehole construction materials

3.1 Casing

Steel casing is fully specified by the American Petroleum Institute (API) specification 5A, 1984, and mild steel water-well casing by British Standards Institute (BSI) specification BS: 879: 1985. There are no official British standards for other casing materials yet, or for screens and gravel packs, but manufacturer's specifications are available. Other national standard specifications for well casing are available. The casing chosen for water wells can be plastic, GRP (glass-reinforced plastic or fibre-glass), or steel.

3.1.1 Plastic casing

Plastic is widely used in shallow aquifers because it is cheap and corrosion free. PVC has been used for production well casing but its use is not widespread in the UK. ABS (acrylonitrile, butadiene and styrene) and, more recently, polyolefin and polypropylene are the most common plastics used for well casing in the UK. The ABS, polyolefin or polypropylene casing is square-threaded and flush-jointed to

BS 879: 1985 and available in 2·9 and 5·8 m lengths in the UK. Heavy-duty, flush-jointed PVC well casing suitable for production wells is available, but PVC pipe with flange and spigot joints is commonly used as casing for observation boreholes. The joints in the PVC well casing are slightly belled to allow threading to Whitworth or trapezoidal designs.

3.1.2 Glass-reinforced plastic (GRP) casing

GRP well casing is relatively strong and corrosion resistant. Lengths are joined by special push-fit lockable joints and collars which have smooth profiles but do increase the OD of the casing (Fig. 3.1). The GRP well casing, when it came out in the 1960's had distinct advantages over plastics in strength and over steel in corrosion resistance and cost, but now these advantages have been eroded, particularly with respect to the improved plastics.

The plastic and GRP casing is more fragile than steel. Threads can be destroyed quite easily by abrasion and the casing cracked by shocks. These

The field guide to water wells and boreholes

Fig. 3.1 Casing connection methods.

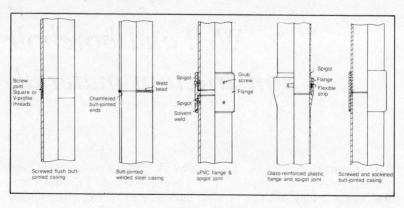

casings also deform more easily than steel and it has been known for casing to be irreversibly crushed by external hydrostatic pressures during well development when excessive drawdowns can occur. The casing should be able to withstand the maximum hydraulic load to which it is likely to be subjected, that is, about 10 kPa for each metre the casing extends below the water table. The collapse resistance of casing depends on the material, the method of manufacture, and the diameter and wall thickness of the casing. There are so many variables affecting the strength of the casing that the well designer is advised to obtain data

from the casing manufacturers for each casing application. The relative strengths of casing made from the most common materials, however, can be judged from the data for 250 mm nominal diameter casing given in Table 3.1. The manufacturing and design details of screens are far more complex than for casing, so it is difficult to obtain comparable meaningful data on their collapse strengths.

Indications of the relative strengths of different casing materials are given in Tables 3.2 and 3.3. The weights of the various types of casing are shown on Fig. 3.2, where the relative lightness of the GRP and plastics compared with

Table 3.1 Casing collapse strengths

Casing material	Casing wall thickness	Collapse strength
PVC (DIN 4925)	12·4 mm	660 kPa
Polypropylene	12·7 mm	690 kPa
GRP	6·0 mm	690 kPa
Mild steel	9·4 mm	11·1 MPa

steel casing is clear. There is little reason not to use the lightweight plastic casings for pump chambers in shallow wells, but it may be wise to use steel for the top length to increase the resistance to surface shocks while pumps are being installed.

Table 3.2 Joint stripping load of plastic casing

	Nominal Internal Diameter (mm)					
	50	*100*	*150*	*200*	*250*	*300*
ABS	1500	2500	5500	6000	10,000	11,500
Polyolefin	1000	2450	5200	5650	9350	10,750
Polypropylene	1000	3600	6200	7600	11,600	14,200
GRP—thin wall			5400	8600	10,000	
GRP—thick wall		11,350	13,600	16,350		

Approximate load (in tension) in kilogram force which will disrupt a string of casing.

Table 3.3 Tensile strength of casing material

	Minimum tensile strength (MPa)
Steel	
After API Spec. 5A, 1984.	
Grade H.40	414
Grade J.55	517
Grade K.55*	655
Grade N.80*	689
After BSI Spec. BS:879:1965	
Grade 26†	402
Grade 27†	417
Grade 35†	541
Plastic	
PVC	55 at 20°‡
ABS	32 at 23°C‡
Polyolefin	30 at 23°C‡
Polypropylene	36 at 23°C‡

* Not normally used for water-well casing
† Grade No. is minimum tensile strength in tons/in^2
‡ Figures approximate but representative of commercial well casing

Fig. 3.2 Casing weights.

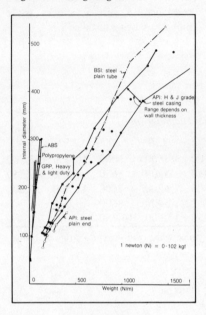

3.1.3 Steel casing

The traditional, and still the most robust casing, is mild steel tubing. Cheap steel tube, ranging from galvanized tin ducting to spiral-weld pipe, is commonly used as casing. However, for a water well, where a guaranteed long life is important, specified steel pipe should be used wherever possible. In the Middle East, casing is commonly specified according to API specifications because water-well drilling developed from oil drilling, but in the UK the BSI specification should be used.

Casing steel is made in several grades (Table 3.3) and the casing is made in several weights (Fig. 3.2). The type of casing has to be chosen to suit local conditions; there is no universal rule, but heavy, high-grade steel casing is for use in the deepest wells while light, low-grade steel casing is used for 'normal' shallow water wells. The definition of a 'shallow' or 'deep' water well will vary with the hydrogeological conditions considered to be standard within an area. In this book, however, it is suggested that water wells up to 200 m deep are shallow, while those over 200 m deep should be considered deep.

Ordinary steel casing is required by BS: 879: 1985 to be coated inside and out with bitumen. Other coatings to protect steel casing from corrosive, aggressive waters have been developed, including a durable hard polyamide. Such coatings are effective until they are broken, so it is extremely important when handling coated-casing to avoid scratching or scraping it. Special rubber-faced tools have been developed for installing such casing. Stainless-steel casing is the best defence against corrosion but it is too expensive to use apart from in exceptional cases.

3.2 Gravel pack design

3.2.1 Natural gravel pack

A natural gravel pack is produced by the development of the aquifer formation itself. Development techniques (Section 8.3) are used to draw the finer fraction of the unconsolidated aquifer through the screen, leaving behind a stable envelope of the coarser, and therefore more permeable, material of the aquifer.

An aquifer is suitable for the development of a natural pack if it is coarse-grained and ill sorted. The grain size and sorting of a sediment are illustrated by a grain-size distribution curve as in Fig. 3.3. The derivation of such a curve is explained in Appendix V, but an important notation from a grain-size distribution curve—used in gravel pack and screen design—is the D number. This is related to the grain size in such a manner that the D40 size, for example, is the sieve mesh diameter through which 40% of the aquifer material will pass. An ill-sorted aquifer has a sorting or uniformity coefficient (D60/D10) of more than about 2·5.

Grain-size distribution curves of a well-sorted and an ill-sorted aquifer are shown on Fig. 3.3. The slot size recommended for the screen to develop a natural pack in the ill-sorted aquifer would be the D40 size, that is, 0·8 mm. This D40 criterion should be used to choose a screen slot size, except when it would dictate a very narrow slot. Extremely fine slots are available with wire-wound screens, but they are susceptible to blockage by corrosion or incrustation products, so it is recommended that a natural pack is not considered for an aquifer when the D40 criterion would dictate a slot width of less than 0·5 mm. It is also recommended that a natural gravel pack should not be considered for an aquifer where the sorting coefficient is much less than 3. In a well-sorted sand (Fig. 3.3) there is so little difference between the D40 and D80 sizes that a slight mismatch of the slot size could lead to a

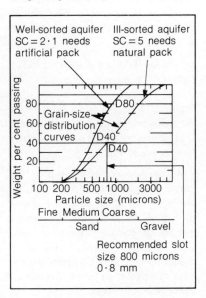

Fig. 3.3 Grain-size distribution curves for gravel pack design.

Well-sorted aquifer
SC = 2·1 needs
artificial pack

Ill-sorted aquifer
SC = 5 needs
natural pack

Weight per cent passing

Grain-size distribution curves

D80

D40

D40

100 200   500 1000   3000
Particle size (microns)
Fine Medium Coarse
Sand   Gravel

Recommended slot size 800 microns 0·8 mm

sand-pumping well; an artificial pack would avoid this danger.

3.2.2 Artificial gravel pack

Artificial gravel packs are used where the aquifer material is very fine, well-sorted or laminated and heterogeneous. Their main advantage is that, because the pack material is coarser than the formation, screens with larger slot sizes can be used. Many designs of gravel pack have been formulated, relating the grain-size distributions in the aquifer and pack (Hunter Blair, 1968; Bakiewicz *et al.*, 1985). The basis of the designs is to select a gravel size distribution that will retain the

33

finer aquifer material, and most are based on the work by Terzaghi & Peck (1948) on filter design. Terzaghi & Peck's suggestion for a filter design can be expressed as:

D15 filter/D85 aquifer <4< D15 filter/D15 aquifer

Where D15 aquifer is the D15 size of the coarsest layer of the formation and D85 aquifer is the D85 size of the finest layer of the formation.

A common consensus is that the gravel pack grading curve should be similar to that of the aquifer, and that the values on each curve for similar 'percentage passing' vary by a constant, commonly about 5 (Fig. 3.4). Bakiewicz *et al.* (1985) suggest that the filter design should be an envelope illustrating tolerances in the gravel make-up.

The concept of a design envelope for the gravel pack design is believed to be more practical than a rigid formula. It is recommended here, therefore, that an artificial gravel pack grain-size distribution should be similar to that of the aquifer being screened, and should lie within an envelope defined by four and six time the aquifer grain size.

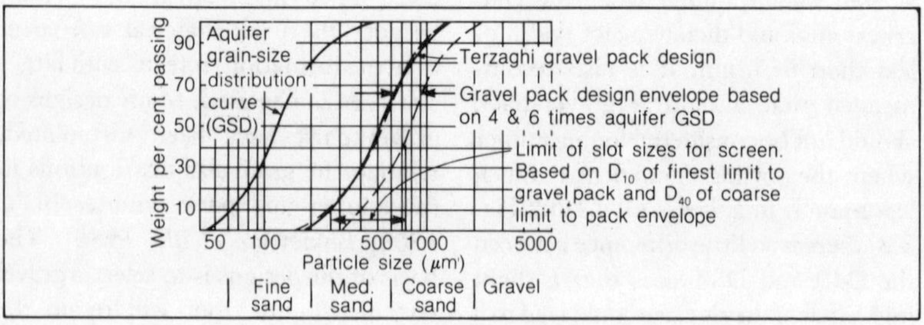

A gravel pack only a few mm thick will retain the formation in undisturbed conditions, but intentional disturbance of the pack and formations during well development means that a relatively thick pack is needed to avoid formation pumping. A thick pack is also able to cope with unintentional thinning of the pack in cases where the screen is not adequately centred. It is recommended that the thickness of the pack should not be less than 75 mm or greater than 150 mm.

A gravel pack less than 50 mm thick is merely a formation stabilizer, acting to support the formation but not acting as a filter. Gravel packs much greater than 150 mm in thickness may create difficulties in development, particularly if mud-flush rotary drilling has been used and a mud cake has to be removed. Also the thicker the pack, the greater will be the drilled diameter of the well and therefore the cost.

Artificial gravel packs, graded in grain size from the coarsest next to the screen to the finest next to the aquifer, have been proposed, but such a design is considered impractical for field installation. In a 'normal' graded pack, installation by tremie pipe is recommended to avoid the different grain

Fig. 3.4 Artificial gravel pack designs.

sizes in the pack settling at different rates—leading to lamination in the pack, and to ensure the gravel completely fills the annulus. Where a thick series of aquifers of different grain sizes is being packed, the pack against each aquifer can be tailored to that aquifer. The accurate emplacement of such a pack is extremely difficult because of pack settlement during subsequent development. The screen and pack for a fine aquifer in such a situation have to extend to at least one metre into the adjacent coarser aquifers, to avoid sand pumping.

The material used for a gravel pack should be natural, sub-rounded siliceous sand or gravel. Ferruginous sand or limestone gravel should not be used, because solution and reprecipitation of iron or calcium salts from the pack are likely to cause problems.

Screens are available with a resin-bonded artificial gravel pack already attached (Fig. 3.3). This screen will act as a filter, but cannot be developed and is susceptible to damage and blockage during installation.

3.3 Screen

The choice of a particular screen type for a water well will depend on a combination of factors: the strength and corrosion resistance, the slot design and the open area (that is, the proportion of a screen face made up of open slots).

The first two factors are dependent on the materials used for screen manufacture—which include plastics, GRP, steels and various alloys. Mild steel screens are susceptible to corrosion and incrustation, while stainless-steel screens are corrosion resistant, strong and expensive. Copper-based alloys are not commonly used for screens.

The perforations in well screens, although normally called slots, are very varied in design and methods of production (Fig. 3.5). They include slots cut by an oxyacetylene torch or sawn in blank casing; bridge and louvre slots, pressed from steel plate and later rolled and welded into tubing; spiral-wound (or wire-wound) screens where a wedge-shaped wire is wrapped in a continuous spiral around a cage of rods and welded to each rod in turn; and perforated pipe base, wrapped with a sleeve of plastic mesh to act as a filter.

The torch-cut slots in mild steel casing are not recommended for water wells because the width of the slots cannot be accurately controlled and the open area of the screen rarely exceeds 2 or 3%. Saw-cut slots are common in plastic or GRP screens and are adequate where slots of 1 mm or more are required. The open area of factory-made plastic screens commonly exceeds 10%. PVC well casing of 200 mm or greater nominal diameter can be provided with slots as fine as 0·5 mm but for any plastic casing, even at a slot width of 1 mm, the ratio of the slot width to screen-wall thickness is so low that the chance of slots being blocked by fine aquifer material must be high.

The bridge and louvre slot screens are of strong construction and can have a high open area. The slot profiles

Fig. 3.5 Commercial screen types.

Mild steel bridge slot

Wire wound

Plastic sawn slot

GRP sawn slot

Resin- bonded pack on PVC screen

Stainless-steel bridge slot

are knife-edged and are not at right angles to the screen face, so they are not as subject to blockage as saw-cut slots (Fig. 3.6). Both types of pressed screens are available in mild and stainless steels. The spiral-wound screens are available in stainless steel, plastic and GRP. The spiral-wound screens are designed to give very high open areas and avoid slot blockage. The 'V' profile of the slots, and the slot widths can be reproduced evenly and accurately to smaller widths than any

other screen type. This type of screen is probably the most suitable for well development and redevelopment (Chapter 8)

Where it is necessary to maximize the yield from an aquifer of limited thickness, stainless-steel wire-wound screen is probably the best, because of its high open area. The bridge slot and louvre slot screens are particularly suitable where strength is required—in deep aquifers or in places where shock loads may be expected. The main ad-

Fig. 3.6   Screen slot profiles.

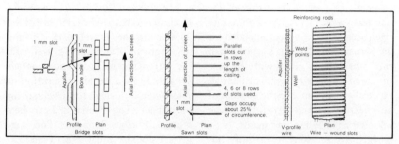

vantages of plastic or GRP screens are their lightness and relative cheapness, but their structural weakness generally restricts their potential use to the shallower water wells. Perforated pipe wrapped with filter fabric material is suitable for observation boreholes but not for high-yielding water wells. It does not allow aquifer development or well rehabilitation and may be susceptible to clogging by clays. It could, however, be satisfactory for water wells in coarse, clean aquifers.

The open area of a screen governs the rate at which water can enter the well. Intuitively one would expect that the higher the open area, the freer the flow into the well, and that those screens with the highest open area would have the best performance. Recent work, however, has suggested that this may not be entirely true. The open cross-sectional area of cubic-packed spheres is 21·5%, while that of rhombic-packed spheres is 9·3% (Nold, 1980, p. 75) and, as the latter is the more stable packing, it is likely that a well-sorted clean aquifer or gravel pack will be much closer to rhombic packing than cubic. The effective open area of an aquifer, therefore, is ex-

pected to be close to 10% and unlikely to be more than 15% (Fig. 3.7). Once the open area of a screen has reached the open area of the aquifer, it is the open area of the aquifer that will govern the flow into the well—extra screen open area will have no effect.

Fig. 3.7 Section of packed spheres, showing open area.

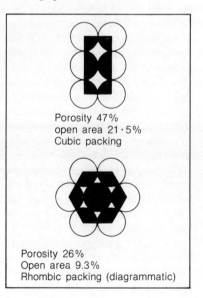

Porosity 47%
open area 21·5%
Cubic packing

Porosity 26%
Open area 9.3%
Rhombic packing (diagrammatic)

37

The open area of a screen should be as great, or greater than that of the aquifer in which it is set, so a minimum open area of 10% is recommended for any screen. In practice, a proportion of the slot open area will become obscured by pack or aquifer material, especially with time, so the concept of an 'effective' screen open area has been adopted as a design parameter. Little work has been done to quantify this term, but, possibly, only 50% of the initial open area of a screen may be effective. This should be allowed for when designing the well.

An important design criterion related to the open area of a screen is the 'entrance velocity'. This is the discharge rate of the well, divided by the effective open area of the screen, and is the average velocity of the water passing through the screen. It is widely accepted that the entrance velocity of the water should be kept below a certain value: 1 ft/min, 3 cm/s or a range of values from 1 to 6 cm/s (Walton, 1962, p.29), otherwise the well will suffer from incrustation, excessive well losses and other destabilizing effects due to turbulent flow conditions. Again, little work has been done to accurately quantify this parameter, but its justification appears to be based on 'experience'. When applying the screen entrance-velocity design criterion it must be remembered that the aquifer is unlikely to be homogeneous, and that entrance velocities opposite zones of high hydraulic conductivity will be much higher than those elsewhere. The open area used to calculate the velocity should be the lesser value of either the open area of the aquifer or the effective open area of the screen.

The width of slots in a screen should be chosen to match the grain size of the aquifer or the gravel pack. In the development of a natural gravel pack the screen should have wide enough slots to allow a certain proportion of fine material through for development of the pack. The criteria used vary between authorities: Johnson (1966) suggests a slot size equal to between D50 and D60 of the aquifer material for a homogeneous aquifer, while Campbell & Lehr (1973) suggest a size between D30 and D50 for a similar aquifer. Johnson suggests that the more conservative D50 size be used where the groundwater is aggressive, but that the criterion may be relaxed in coarse-grained aquifers so that the slot size is between D50 and D70 of the aquifer. Campbell & Lehr use their conservative D30 when the screened section is overlain by fine material and the D50 slot size when it is overlain by coarse material. The authors relax their criteria when screening heterogeneous aquifers to D70 when the screened section is overlain by coarse material, and to D50 when it is overlain by fine material.

It is recommended here that the choice of slot size should be based in most cases on the rather conservative average D40 (Fig. 3.3) of the grain-size analyses of at least six aquifer samples. In coarse or ill-sorted aquifers a more relaxed criterion can be adopted, and slot sizes of D50 and over of the aquifer could be tolerated. The slot choice for a screen in a very heterogeneous aquifer should be based on analyses of the finer parts of the aquifer, not on 'average' aquifer material. In a layered aquifer the screen can be of multiple construction with a slot size to match each layer. In such a construction the fine screens have to extend about a metre into the coarse aquifers to avoid sand pumping, and only in the thicker aquifers can sampling be adequate enough for this to be done.

The size of the slots in artificially gravel-packed wells is generally recommended to be D10 of the pack material (Johnson, 1966; Hunter Blair, 1968). More recently, Bakiewicz *et al.* (1985)

recommended for gravel-packed wells in Pakistan that the slot width should be less than half the D85 of the gravel. This criterion incorporates a safety factor of 2, but even so, would indicate a slot size of around D40 of the gravel. It is suggested here, therefore, that slot sizes up to D40 of the pack material could be used generally, but that the more conservative D10 rule could be used to choose the slot width when a very uniform pack is being installed. Table 3.4 summarizes gravel pack and screen selection.

Table 3.4 Summary of gravel pack and screen selection

---

1. Is the aquifer:
   a) Crystalline? Yes—then no screen or pack is required. Formation stabilizer may be needed (p. 30).
   b) Consolidated? Yes—then screen and pack are usually not required (p.14). If needed for multiple aquifer system (p.18) use formation stabilizer or follow criteria for unconsolidated aquifer.
   c) Unconsolidated aquifer? Yes—then screen and gravel pack needed.

Then plot grain-size distribution (GSD) curves for samples of aquifer material (Appendix A.V).

2. Is the aquifer heterogeneous (sorting coefficient > 2.5)?
   Yes—then a natural gravel pack can be developed (p.29). The screen slot width should be the average D40 of the aquifer samples (Fig. 3.3). No—then an artificial gravel pack is needed (p.30). Design the pack by plotting two curves parallel to the GSD curve of the finest aquifer sample but four and six times coarser than that curve (Fig. 3.4). The GSD curve of the gravel pack used should lie between the two new curves. The screen slot width should be between the D10 and D40 of the gravel pack.

3. The screen diameter will have been decided during the structural design (p.17). The choice of screen material and slot design will depend on site conditions:
   a) Is the aquifer very thick? Yes—then a long screen with limited (but > 10%) open area may be chosen.

Table 3.4 *continued*

---

b) Is the aquifer thin? Yes—then a screen with a very high open area should be chosen.

c) Is the aquifer over 200 m deep? Yes—then a strong screen must be chosen: mild or stainless steel.

d) Is the groundwater corrosive? Yes—then plastic, GRP, stainless steel or coated screens should be used.
No—then mild steel screens can be used.

e) Is the groundwater encrusting? Yes—then screen with a high open area should be used to reduce entrance velocities (p.33).

4

Well and borehole construction

The drilling methods and machinery used for well and borehole construction are too varied to be covered comprehensively in this manual. This chapter is intended to provide sufficient knowledge of the basic principles of the main drilling techniques, for an engineer or hydrogeologist to prepare adequately or supervise a drilling contract and to recognize equipment available in depots or brought on to the site.

The person letting a drilling contract will have to assess the suitability of various drilling organizations to undertake the work. Important points to be considered in such an assessment are summarized in Table 4.1.

Table 4.1 Points to consider in selecting a drilling contractor

---

1. **Drilling organization's history**
    Has the company a local or national reputatation for good work?
    Can they provide details of similar drilling works that they have completed recently?
    Do they belong to a professional organization? (In the U.K for example, the British Drilling Association).
    Are they on a list of approved constractors for an organization employing drilling contractors?

2. **Visit to drillers depot**
    Have they adequate equipment to do the work?
    Is the equipment in good repair and well maintained?
    Has the contractor maintenance and back-up facilities in case of plant breakdown?
    Is the depot close to the proposed drilling site so that back-up can be efficient?
    Is the depot kept in a clean and workman-like state?

Drilling techniques vary, but most of them can be classified as either percussion or rotary techniques. Percussion drilling is used most often at shallow depths, while rotary methods predominate in the construction of deep boreholes or wells. Hand digging of wells is discussed seperately.

The driller must keep a daily record of work on site to make up a 'driller's log'. The daily record (Tables 4.2 and 4.3) should contain the following information with respect to each borehole or well:

1. Site location.
2. Number of borehole or well within the contract.
3. Date.
4. Method of boring and rig used.
5. Depth of hole.
6. Type, length and diameter of casing.
7. Type, length and diameter of screen.
8. Length and diameter of open hole.
9. Water levels, with details of any fluctuations.
10. Description of each stratum encountered.
11. Depth below ground of any change in lithology.
12. Sample depth, type and characteristics.
13. Any other useful information, for example, the rate of drilling.

4.1 Percussion drilling

The main features of a percussion rig are shown in Fig. 4.1. A string of heavy cutting tools is suspended on a cable which passes over a sheave mounted on a mast, beneath a sheave on the free end of a spudding arm, over a sheave at the base of the spudding arm, and is then wound on a heavy-duty winch. This use of the steel cable gives the technique its American name of 'cable-tool drilling'.

Power is supplied by a diesel engine. The whole rig can be mounted on the back of a truck or trailer, and is quite a mobile unit.

The cable is a non-pre-formed, left-hand lay, steel-wire rope. The left-hand lay of the cable (Fig. 4.2) tends to impart a slight rotation to the tool string and to tighten the right-hand threaded joints of the string.

The tool string generally includes the following units: rope socket or swivel, jars, sinker bar and drill bit (Fig. 4.3). The rope socket has an internal mandrel which allows the tool string to swivel and so prevent overtwisting. During drilling, the cable rotates the tool string slightly, but when the string has rotated sufficiently for torque in the cable to have built up, the mandrel on the rope socket allows the tool string to swivel back. Jars are interlocked sliding bars, which allow a free stroke of about 150 mm. They are not used for drilling, merely for releasing tools by upward jarring if the drill bit becomes trapped. The sinker bar, or drill stem, is a solid steel rod used to give extra weight above the bit, and to improve the verticality and straightness of the hole. The units of a tool string are connected by standard taper thread cable-tool joints.

Table 4.2 Daily borehole record for percussion drilling

L.C. Boring Ltd.		Site		Page			
		Borehole no.		Date			
Type of rig		Type of bit					
Depth (m)	Formation description	Depth to water (m)	Sample*	Casing Diam. (mm)	Depth (m)	Screen Diam. (mm)	Depth (m)
Groundwater level: start of shift time				End of shift time			
Chiselling: times							
Standing time: times							
Remarks:							
						Foreman	

*Samples: D = disturbed, U = undisturbed (U100 or core).

43

The field guide to water wells and boreholes

Table 4.3 Daily borehole record for rotary drilling

L.C. Boring Ltd.		Site	Page
		Rig	Mud Pump

		Type of drilling		
Depth (m)	Formation description	Drilling fluid type, col, visc (sec.)	Bit type and diam. (mm)	Sample* D C%

Core barrel details:

Casing record	Type Diam. (mm) Top (m) Bottom (m) Date installed

Screen record	Type Diam. (mm) Top (m) Bottom (m) Date installed

Remarks:

Foreman

*Samples: D = disturbed, C = Core, % = Percentage Recovery.

Fig. 4.1 Percussion drilling rig for water wells.

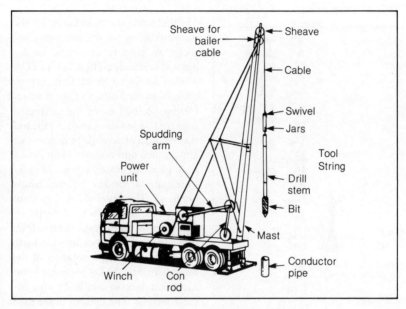

Fig. 4.2 Left lay cable for percussion rig.

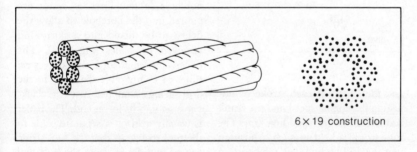

6 × 19 construction

The drilling of a borehole is started by installing a short, large diameter conductor or start pipe in the ground, either by digging or by drilling a pit 1–2 m deep. This conductor pipe is to prevent the surface material beneath the rig from collapsing into the hole and to guide the tool in the initial drilling. The tool string is assembled and lowered into the conductor pipe, and then drilling begins. The driller can vary the number of strokes per minute

45

Fig. 4.3 Tool string for percussion drilling in hard rock.

4.1.1 Hard rock drilling

The tool string shown in Fig. 4.3 is the standard string for drilling such hard rocks as granite, limestone or indurated sandstone. The drill bit will be a heavy solid-steel chisel. Drilling proceeds from the bottom of the conductor pipe to the base of the weathered zone, or to a predetermined depth, and then a length of permanent casing is set and grouted into position. This casing must be set accurately vertical in order to ensure verticality of the whole borehole. The borehole is then continued open-hole to its total depth.

The action of the heavy chisel is to fracture and pound the rock into sand-size fragments. The slight rotation of the bit imparted by the lay of the cable ensures that the hole is drilled with a circular section. The bottom of the hole becomes filled with rock debris which has to be removed periodically to allow drilling to progress freely. In a dry borehole, a few litres of water are poured into the borehole to allow the debris to be mixed into a slurry and removed by bailer (Fig. 4.3). The bailer is a length of heavy duty steel tube, of comparable diameter to the drill string, equipped with a clack-valve set into its lower end. The bailer is usually held on a separate cable to the tool string, so that the tool string can be withdrawn from the hole and the bailer lowered with minimum interuption of the operations. Agitation of the bailer at the bottom of the borehole ensures that the drilling debris is held in suspension and is forced into the bailer through the clack-valve. The slurry is removed from the

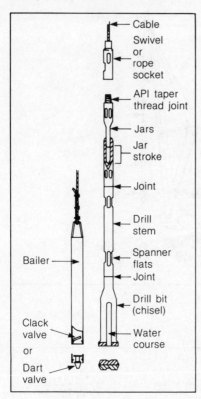

Fig. 4.3 Tool string for percussion drilling in hard rock.

Cable

Swivel or rope socket

API taper thread joint

Jars

Jar stroke

Joint

Drill stem

Spanner flats

Joint

Drill bit (chisel)

Water course

Bailer

Clack valve or Dart valve

and the length of each stroke by adjusting the engine speed and the crank connection on the spudding arm. The tool string is held in such a position that the drill bit will strike the bottom of the hole sharply, and the cable is fed out at such a rate that this position is maintained.

The actual procedure of drilling will depend on the formation to be drilled, but can be illustrated by the procedures for hard rock and soft unstable rock.

hole and the operations repeated until the hole is clean; drilling then continues. Samples of the slurry are taken from the bailer to produce a lithological log (Section 7.2).

The addition of water is not necessary for drilling or bailing once groundwater has been struck. Below the water table the bailer contents can be used for analysis to get an indication of the groundwater quality. The driller should note when water is struck, and measure the water level at the start and end of each shift. These data can give valuable information on the groundwater regime and the borehole productivity.

4.1.2 Drilling in soft, unstable formations

The tool string used for drilling in soft, unstable formations is commonly simpler than for hard formations. The most common tool is a steel tube or shell with a cutting shoe on the bottom (Fig. 4.4). The shell may be a plain tube or have windows cut in the side to help sample removal. The shell may be used alone or, if extra weight is needed, a sinker bar may be incorporated in the body of the shell. The top end of the shell is open and it is very similar in design and appearance to a normal bailer.

The cutting shoes differ in design in order to cope with different lithologies. With sand, the shoe may have a serrated or smooth edge to chop the formation, but a clack valve will be set inside the shell, just behind the shoe, to retain the sand as it is cut. Clay tends to

Fig. 4.4 Shell for drilling soft formations.

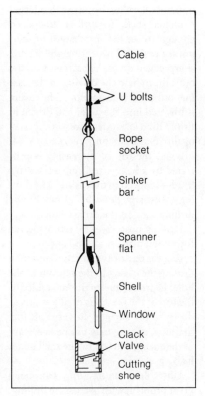

Cable

U bolts

Rope socket

Sinker bar

Spanner flat

Shell

Window

Clack Valve

Cutting shoe

be more coherent than sand, and, with stiff clay, a sharp edged cutting shoe with thin chisel blades set across its aperture may be used. The chisels cut the clay into pieces as it enters the shell and help to retain the clay in the shell.

A major difference between unstable formations and hard rocks is that the former need support during drilling and this means that temporary casing has to be used. The temporary casing has to be of sufficient diameter to allow

47

the permanent casing/screen string to pass inside it on completion of the borehole. Drilling begins with a large-diameter shell, to drill a hole deep enough to set the first length of temporary casing. This casing will have a sharp-edged drive shoe screwed on the bottom, to help the driving of the casing into the soft formation. The casing is lowered into the hole, and driven in firmly for a short distance with its verticality being carefully checked. While driving, the top of the casing is protected by a heavy steel ring—the drive head—screwed to the casing. The driving is usually done by blows of the drilling tool on the drive head.

Drilling progresses by removing the formation from inside and ahead of the casing for one or two metres, and then driving the casing to the bottom of the hole. Extra casing lengths are added as the top of the previous casing is driven down to the surface. In very soft formations, driving may not be necessary as the casing will follow the shell under its own weight.

Unstable running sand can be a great problem with percussion drilling, because the reciprocating action of the shell inside the casing can build up a suction which pulls the sand up inside the casing. Remedies for running sand include building up a hydraulic head in the casing by filling it with water, adding drilling mud to stabilize the sand, driving casing past the unstable zone, or grouting off the running sand.

Drilling continues until the total design depth is reached or until the casing cannot be driven further. In the latter case, a second string of temporary casing has to be installed inside the first

and drilling continues at a smaller diameter. The initial well design on which a drilling contract is based must always make allowance for this contingency, because the permanent casing has to be able to pass inside the smallest temporary casing.

On reaching the total depth, the design of the permanent casing/screen string is finalized and the string installed. The screen must be located accurately in the aquifer, and so the formation samples must be taken carefully and described accurately (Section 7.2). The depth of samples can be measured fairly accurately with percussion drilling, but geophysical logging is still recommended to obtain precise depths of the aquifer. The log must be a natural gamma ray log or one of the radioactive logs (Section 7.3) because the temporary casing in the hole prevents any other logs being used. After logging, the casing/screen string is installed and the temporary casing removed. The drive head is screwed on the top of the casing, and a continuous tension is put on the head through the drill cable. The temporary casing is commonly reluctant to move and has to be started by upward jarring, either by using the shell itself or a set of jars in the tool string. In extreme cases, jacks or a hydraulic vibrator may have to be used to break the casing free.

In a borehole or well where a gravel pack has to be installed, the formation samples must be analysed in order to design the gravel pack and to decide on screen design (Chapter 3). The analyses can be done in the laboratory, but preferably are done on-site to save time; this is important in isolated locations (Appendix A.V). The permanent casing/screen string is installed inside the temporary casing and the gravel pack poured into the annulus through a tremie pipe. The gravel or sand may have to be flushed down the tremie pipe with water, and the casing should be vibrated to encourage settlement of the pack and avoid gravel bridging. Great care has to be taken to withdraw the temporary casing just ahead of the pack, otherwise a sand lock between the screen and the temporary casing can occur, which binds the two together, and then the screen will be withdrawn with the casing. The depth to the top of the gravel must be checked frequently with a plump-bob.

The greatest strain on a percussion rig is during the withdrawal of the temporary casing. The size and robustness of a rig and its fittings must be sufficient to cope with the pull it has to exert. This is particularly important in deep, large-diameter boreholes that are drilled in formations which may collapse around and grip the temporary casing.

A great number of boreholes, particulary those drilled for site investigations or for shallow explorations, are small structures. A percussion rig has been developed specifically for this kind of borehole, with an emphasis on mobility and lightness. This rig is based on a tripod instead of a single mast (Fig. 4.5) and can be collapsed to be towed behind a small truck or site vehicle. The drilling operations are identical to those with a traditional rig but there is no spudding arm, and the reciprocating action of the tool string is

Fig. 4.5 Small percussion rig.

achieved by direct operation of the cable winch.

The driller will keep a record of work done on the site and a typical daily record sheet for percussion drilling is shown on Table 4.2.

4.2 Rotary drilling

Rotary drilling frequently overcomes the problem of having to use temporary casing, by using the hydrostatic pressure of circulating drilling fluids to support the borehole walls. This use of drilling fluids enables boreholes to be drilled to much greater depths than can be achieved by percussion rigs. Rotary techniques can be divided into two main types—direct circulation and reverse circulation, depending on the method used to circulate the drilling fluid.

4.2.1 Direct circulation

A typical design of a direct circulation rotary rig for drilling water wells is shown on Fig. 4.6. The body of the rig comprises a floor on which is mounted a diesel-engined power unit, a mud pump for circulating the fluid, a winch for raising or lowering the drill string and a mast from which the drill string is suspended. The drill string is made up of lengths of heavy-duty steel tubing or drill pipe, with the drill bit assembly attached to the bottom. All members of the drill string are connected by standard API taper thread joints.

The drilling fluid or mud is mixed in a mud pit or mud tank and is pumped by the mud-pump through the kelly hose to a water swivel at the top of the kelly. This swivel is the unit from which the entire drilling string is suspended, and allows the mud to pass while the drill string rotates. The mud passes down the drill string to the bit which it leaves by ports in the bit faces, and then returns up the annulus around the drill string to the mud pits. A mud pit often has at least two chambers, the first and largest allows cuttings to settle from the mud, before it passes to the second chamber which acts as a sump for the mud-pump. The first chamber, the settlement pit, should be twice as long as it is wide, and be at least three times the volume of the hole to be drilled.

Fig. 4.6 Direct-circulation rotary drilling rig.

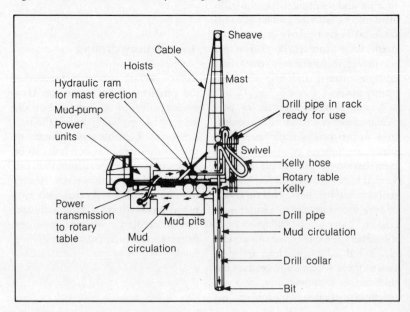

The drill pipe is normally of circular section, but the top length, the kelly, usually has a square or hexagonal external section. The kelly passes through a similar shaped hole in the rotary table at the foot of the drill mast. This enables rotary power drive to be transmitted from the main power unit, through the rotary table, to the kelly and so to the entire drill string.

The drill bit assembly commonly consists of the bit itself with, immediately above, a length of large-diameter, very heavy drill pipe called a drill collar. The collar is designed to give weight to the drill string, improve its stability and hence the hole verticality, as well as decreasing the annulus around the string, and so increasing the velocity of mud flow away from the bit.

Drilling begins with the installation of a length of conductor pipe to prevent erosion of the surface by the mud flow. The water swivel is then hoisted up the mast and the drill bit assembly screwed on to the kelly. The bit is lowered into the hole, the kelly clamped into the rotary table, and mud circulation and rotation of the drill stem are begun. The kelly can pass vertically through the rotary table quite freely, and drilling progresses under the weight of the drill string. When the swivel reaches the rotary table, the drill string is held suspended on the hoist, and mud circulation is continued until all the cuttings have been removed from the hole. Circulation and rotation are then stopped, the drill pipe suspended by friction slips in the rotary table and the kelly is unscrewed. A new length of drill pipe is fastened to the

Fig. 4.7 Schematic diagram of top-drive rotary rig.

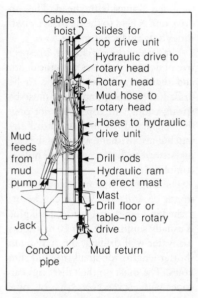

kelly and drill string, the slips removed, and circulation and rotation restored to start drilling again. The momentum of drilling needs to stop only when the final depth is reached or the drill bit needs replacing. The withdrawal of the drill string is a reversal of the drilling procedure.

The use of a rotary table to transmit the rotary drive to the drill string allows a very robust design, suitable for rigs varying from small truck-mounted rigs, to large platform rigs similar to those used in the oil industry. An alternative design, commonly used in rigs for water wells, is the top drive rig (Fig. 4.7). The principles of drilling are identical in the two designs, but in the top drive there is no kelly, the drill pipe being attached directly to the

51

rotary head. Rotary drive is transmitted to the head by a hydraulic motor mounted alongside the head. The top drive unit is held to the rig mast by two slides down which it moves.

The size of the rig chosen, irrespective of design, increases with the depth and diameter of the borehole to be drilled. The hoist and mast must be strong enough to cope with not only the weight and vibration of the drill string, but also with the weight of the casing strings to be set when the hole is completed.

The height of the mast governs the length of drill pipe which can be attached in a single operation. Drill pipe is usually supplied in 3 to 10 m lengths for water-well drilling, but may be in shorter lengths for small rigs and slim holes. The mast on the tallest rigs can cope with several lengths of pipe fastened together—so speeding up operations. The internal diameter of the drill pipe must be large enough to allow mud to pass freely down the drill string, and the outside diameter should be large enough to ensure sufficient velocity in the mud (Appendix A.II.2) passing up the annulus to carry cuttings from the bit to the surface.

The design of drill bit used depends on the formation to be drilled. In soft formations a simple drag bit equipped with hardened blades can be used. The commonest rotary bit is the tricone bit (Fig. 4.8) which has three hardened steel, toothed conical cutters which can rotate on bearings. The drilling fluid passes through ports which are placed to clean and cool the teeth as well as carry away the cuttings. The teeth on the cutters vary in size and number to suit the formation being drilled, small, numerous teeth for hard formations and larger teeth for softer formations. The bit operates by rupturing the rock: by overloading it at the points of the teeth and by tearing or scraping as the cones rotate. Drag and tricone bits break the rock into fragments or cuttings, which are returned to the surface to provide 'disturbed formation samples'.

Samples of undisturbed formation can be obtained as cores of strata (Chapter 5), by using special coring assemblies which comprise a tubular diamond or carbide-studded bit attached to a core barrel which, in turn, is attached to the drill string. The bit cuts a solid rod of formation, over which passes the core barrel where it is held. When the barrel, which is usually 1·5 or 3 m long, is full, the entire drill string is removed from the borehole to retrieve the core from the core barrel.

Fig. 4.8 Tricone drill bit and diamond-faced coring bit.

An alternative system, wire-line core drilling, has been developed to avoid the need to withdraw the drill string while coring. Special wire-line drill pipe, with an internal diameter large enough for the core barrel to pass, is used for the drill string, and when the core barrel is filled, the barrel and core are retrieved by a special bayonet tool lowered on a cable down the inside of the drill string. The bayonet locks into a female head of the core barrel.

Drilling fluids An important factor in direct circulation drilling is the choice of the drilling fluid (Appendix A.II.a). The fluid can be clean water if the formations are hard and competent. Water is the usual fluid for slim hole exploratory drilling for mineral exploration in crystalline rocks. The most common general-purpose drilling fluid for sediments is a mud based on natural benonite clay. The mud fulfils several purposes; it:

1. Removes cuttings from the bit, carries them to the surface, and allows them to settle out in mud pits.
2. Cleans, cools and lubricates the drill bit and drill string.
3. Forms a supportive mud cake on the borehole wall.
4. Exerts a hydrostatic pressure through the mud cake to prevent caving of the formation.
5. Retains cuttings in suspension while the drilling stops to add extra lengths of drill pipe.

The properties of the mud which allows it to fulfil these functions are its velocity, viscosity, density and thixotropy. The velocity of the mud moving up the borehole will depend on the speed and capacity of the mud-pump, and on the cross-section of the annulus between the drill string and the borehole wall. Large cuttings tend to sink through the mud, so its upward velocity has to be greater than that with which the cuttings are sinking. The sinking velocity of a coarse sand fraction in water is about $0 \cdot 15$ m/sec but will be much less in viscous mud. The viscosity of the mud controls the rate at which the cuttings sink; if it is too high, then the cuttings may not settle in the mud pit. This leads to recirculation of the cuttings, which causes excessive wear on the mud-pump and drill bit, and a mixing of cutting samples (Chapter 5). The viscosity of the mud may be increased through the addition of formation clay, or decreased by the influx of waters varying quality. The mud viscosity, therefore, has to be monitored constantly by means of a Marsh funnel. The funnel is filled with one litre of mud, and the time measured in seconds for it to empty; the viscosity is then expressed in seconds. The buoyancy offered by the mud, and hence its ability to carry cuttings, will increase with its density. A good bentonite mud will have a Marsh viscosity of 30–40 seconds and a density of 10 lb/US gallon ($1 \cdot 2$ kg/l) or a specific gravity of $1 \cdot 2$.

The mud is a suspension, partially collodial, of clay in water and, under the hydrostatic pressure of the column of mud in the borehole, water is forced from the suspension into the adjacent formations. The water leaves the clay behind as a layer or cake attached to the borehole wall. The filtration will be greatest adjacent to the more permeable formations (Fig. 7.5). The hydraulic pressure exerted by the mud column depends on the mud density and, in severely caving formations, the mud density can be increased by the addition of heavy minerals such as barytes. Care must be taken, however, to avoid execessive mud cake build up, because if it becomes too thick it can reduce the diameter of the borehole sufficiently to prevent withdrawal of the drill string.

Bentonite clay is thixotropic and this ability to gel when not disturbed is important in holding cuttings in suspension when mud circulation stops. In a 'thin' mud, there is a danger that cuttings will settle behind the bit assembly and lock it down the borehole.

Bentonite-based mud has some important disadvantages for drilling water wells. The mud cake can be securely keyed to the porous formations so that it is difficut to remove during well development (Section 8.2). When a gravel pack has been installed in front of the mud cake, well development may be almost impossible. The cake can also prevent a good grout-key when casing is being grouted in. In the build-up of the mud cake, a filtrate of drilling water will invade the formations and fine sediment carried by this filtrate can severely reduce the formation permeability. The removal of this filtrate and sediment by development is difficult and rarely completely successful. In order to avoid these problems, alternatives to bentonite have been developed, the most important being organic polymers, foam and air.

Organic polymers When mixed with water, organic polymers form a viscous fluid with many of the characteristics of bentonite-based mud. However, after a certain time, the polymer mud breaks down to a low-viscosity fluid which can be removed far more easily by well development. The natural life of the polymer mud varies with the polymer and local conditions. One commonly used organic polymer has a life of about four days, but breakdown can occur much more quickly if the drilling water supply is bacteriologically contaminated or if the pH is low, below about pH 4. Stability of the polymer mud can be improved by various proprietry additives. Formaldehyde was the commonest additive but is now replaced, because of its toxicity, by food-grade inhibitors. On completion of a borehole the breakdown of the mud can be accelerated to within a hour, by an addition of chlorine to the drilling fluid.

Organic polymers have obvious advantages over bentonite as a mud base:

1. The controlled breakdown of the drilling mud to a water-like liquid on completion of drilling assists greatly in the removal of the mud cake and in the well development.

2. Less polymer than bentonite is needed to make an equivalent mud.

3. The formation samples recovered are much cleaner. There is no bentonite to remove or distinguish from formation clays, so that lithogical logs and grain-size distribution analyses are more accurate.

There are also some disadvantages of organic polymer muds:

1. The mud condition has to be more closely monitored to avoid unexpected breakdown. The manufacturer's guidelines for use have to be followed closely to ensure optimum performance.
2. The organic polymer can act as a food source for bacteria in a well, and has to be totally removed to avoid subsequent bacterial infection of the well. Again the manufacturer's guidelines have to be followed closely.
3. The polymers solve the problem of mud-cake removal to a large extent, but do not avoid the problem of formation invasion by fine particles from the mud.

A major problem with drilling with either bentonite or polymer-based muds in extremely porous formations or in fissured (karstic) limestones, is the loss of circulation. In karstic limestone, the entire mud circulation can disappear into a major fissure. This results in a loss of formation samples but, more seriously, a loss of lubrication between the drill string and the borehole walls, loss of support for the borehole walls and a need to constantly replenish the mud supply. The problem can sometimes be solved by plugging the zone of lost circulation with a bulky medium such as wood chips, or emplacing grout to grout-off the loss zone. Unfortunately, this may also grout-off productive aquifers and, in the case of large fissures, may merely result in the loss of large volumes of grout.

Foam-based drilling fluids can be used as an alternative to muds where fluid loss is excessive. Proprietary chemicals, akin to domestic detergents, are added to the drilling water, together with compressed air, to give a stable, very thixotropic foam. The mix of foaming agent, air and water used will depend on the agent and the site conditions. The mixture is passed down the drill stem under pressure, but, on release of pressure in any cavities, a stiff foam will form (of the consistency of shaving foam), block the cavities and restore circulation. The foam has sufficient consistency to carry cuttings to the surface where it should break down. If the foam is too stiff or the circulation rate too high, then the foam can build up to be troublesome, particularly on windy days. The area of settlement pits should be large enough to cope with the foam produced.

Compressed air as the circulation medium avoids the use of liquids as a drilling medium altogether and can be very effective for small-diameter boreholes. When drilling observation boreholes for groundwater pollution studies—where it is essential that drilling operations affect the groundwater quality as little as possible—air drilling may be mandatory. The technique, however, does present several problems:

1. The low density of the air means that the return velocities must be high to carry cuttings to the surface. The need for a high velocity means that for large-diameter boreholes, very large capacity com-

pressors may be needed to supply the necessary volume of air (Fig . A.II.1)

2. Air drilling presents few problems in dry boreholes, but below the water table the air pressure must overcome the hydrostatic pressure of the water column as well as be sufficient to air-lift the water and cuttings to the surface.

3. In 'damp' formations, cuttings may stick to the borehole walls above the bit and form a mud collar which can prevent withdrawal of the drill string.

4. In pollution studies, particulary studies of organic pollution, air drilling can pollute the formation with oil derivatives from the compressor. An oil trap on the air line is essential for drilling in these conditions.

4.2.2   Reverse circulation.

The reverse circulation system was developed for drilling large-diameter boreholes in loose formations. The layout of a rig is shown in Fig 4.9. The rig is similar to a direct circulation rig, but the drilling fluid is circulated in the opposite direction and many rig components are larger than on a direct circulation rig. The drill pipe has a much larger internal diameter, usually 150 mm minimum, to provide a large waterway, and tends to be in short (about 3 m) lengths of heavy-duty steel tube with flanged connections. The drill bits used at large diameters are commonly composite bits of variable design, but all have an open end to allow cuttings to enter. One design has rings of conical cutters set on the side of the bit body, with the diameter of the rings increasing away from the tip to ream out the hole progressively to its full diameter. At large diameters, extra weight is needed in the drill string above that of the drill pipes, to effect penetration and maintain stability. For this reason a heavy drill collar, together with large-diameter stabilizers is commonly fitted above the drill bit. The stabilizer, above the collar, has an ID equal to the central waterway, but an OD close to that of the borehole. The stabilizer is designed with a large diameter to keep the bit centralized, but with large channels to allow free passage of the drilling fluid (Fig 4.10).

Rotary drive can be transmitted to the drill stem via a rotary table and kelly as in Fig. 4.9, or by a top-drive mechanism. The rate of rotation is much slower than with direct circulation, but the weight of the drill string leads to rapid drilling. Care must be taken by the driller to avoid excessive weight on the bit, which can lead to build-up of torque in the string and rupture by twisting the drill pipe.

The drilling fluid is usually water which is pulled up the drill pipe by a centrifugal pump commonly aided by airlift. The kelly and drill pipe lengths are short to avoid the need for high suction heads in the system. The water passes through the swivel, is discharged into a large settling pit, and then flows by gravity down the annulus around the drill string to the bit. The flow of water up the drill string is at high velocity, 3–4 m/sec, and is capable of lifting pieces of drilling

Fig 4.9 Reverse circulation rig: diagrammatic.

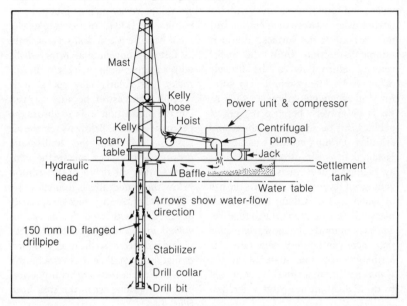

debris of a size close to the waterway diameter. These cuttings, when in water, settle readily on the settlement pit before the 'clean' water returns to the borehole.

Reverse circulation is based on the ability of the borehole walls to be supported by the hydrostatic pressure exerted by the water column and the positive flow of water from the borehole to the formation. The water level in the borehole has to be kept at the surface or, if the water table is shallow, kept above the surface to maintain sufficient hydrostatic head. This means that there is a constant heavy loss of water to the formation and this has to be replenished, usually by tanker. The top-up water required

will increase with the thickness and permeability of formation drilled, but commonly exceeds 50 m^3/h (10,000 gph).

Fig. 4.10 Reverse circulation stabilizers.

The reverse circulation system provides good formation samples with little time delay between their cutting and their arrival at the surface, although sample collection from the high-velocity return flow can be difficult (Chapter 5). The system is very good for drilling coarse sediments such as sands or gravels, because the gravel pebbles can be removed without any grinding. Drilling in such situations is faster than with direct circulation, averaging about 10 m/h. A further advantage of reverse circulation is the use of water as the drilling fluid, which means there is no mud cake to remove. The lack of mud cake, however means that there can be deep penetration of formations by fine material in the water, an action positively encouraged by the hydrostatic head and water loss.

exploration boreholes. In hard, crystalline rocks such as limestone or basalts, the DTH hammer is in its element because it can drill several times as fast as conventional rotary drills with a tricone bit in such rocks. In soft rocks, particularly clay rocks, it is much less successful, because clay can jam the slide action and can absorb the percussive blows. Below water, the air has to overcome the hydrostatic pressure before it can operate the hammer, so that the depth of penetration below the water table is limited.

The DTH hammer bits are specially designed to withstand the shocks involved in the technique. Sharp teeth would be broken, so the most common design for crystalline rocks is a button bit in which the teeth are hemispherical carbide buttons set in the steel head (Fig. 4.11).

4.2.3 Down-the-hole (DTH) hammer drilling

The DTH hammer technique was developed in the quarry and mining industries. The rig and drill string are essentially the same as for direct circulation drilling, but the drill bit assemblies differ. The DTH hammer assembly is a pneumatic hammer, similar in action to a common road drill, in which the compressed air supply operates a slide action to give rapid percussive blows to the bit face as it is rotated by the drill string.

The DTH hammer was originally designed for drilling slim shot holes, but now bits of 200 mm diameter or more are available for water wells and

Fig. 4.11 Button bit for down-the-hole hammer drilling.

4.2.4 Methods of casing and screen installation

The methods of casing and screen installation in rotary-drilled boreholes vary with the depth of the borehole and the ability to install the casing/screen string in one operation.

A borehole for a single string completion is drilled to the design depth and then geophysically logged. The aquifer boundaries, upper and lower, are identified and the casing/screen string is assembled on the surface into lengths suitable for handling by the available rig. As the string is lowered down the hole, each length is joined to the top of the previous one, the joining being by solvent welding, electric welding or screwed joints—depending on the material being used (Fig. 3.1). The casing/screen string is lowered until the screen is opposite the target aquifer. The annulus is then filled either with a gravel pack or formation stabilizer, or the formation is allowed to collapse against the screen during development. This method of installation can be used for single or multiple aquifers; in both cases close control of the depth of the screen(s) must be kept to avoid misplaced screens.

The casing/screen string is held in the centre of the borehole by means of centralizers fixed around the string at 10 to 20 m centres. Centralizers are barrel-shaped cages of spring steel ribs and, apart from keeping the string central in the hole, they serve to guide it down the borehole. Other useful attachments to the casing/screen string, in a borehole drilled in a stable formation, are wall-scratchers which are strapped to the casing at intervals. They are collars of radiating spring-steel wire spokes which help to scrape off the mud cake and improve the keying of grout seals, and to help in the well development (Fig. 8.4).

A casing/screen string where the screen is a smaller diameter than the pump chamber (Fig. 2.9) requires special care in assembly. The two lengths of different diameter are often connected by a conical reducer and it is essential, particularly if this is made on-site, to ensure that the upper and lower sections are coaxial.

The screen in a multiple string completion is part of the last string to be installed. The borehole is drilled for the pump chamber, the pump chamber casing is set and then grouted. Drilling then continues at a reduced diameter, the intermediate casing is set and grouted into position. Drilling then continues through the aquifer(s) at the final diameter to the full depth. The aquifer succession is geophysically logged, and the position and length of the casing and screen string established. The string is assembled on the surface and lowered into the well, attached to a special sub on the drill stem of the drilling rig. When it is adjacent to the aquifer(s) the screen can be set in position or attached to the intermediate casing by several methods. The simplest is to let the screen rest on the bottom of the borehole and be held in position by centralizers, the collapsed formation or the gravel pack. The top of the screen string should extend a few metres up inside the casing above and the annulus between should be left open. With this kind of completion, the gravel pack can be put down the annulus by tremie pipe, but it does allow formation material to be washed into the well. The top of the annulus can be sealed by a flexible sheath of lead or similar malleable material hammered into a conical shape. A custom-built seal is available, where a neoprene seal assembly, attached to the top of the screen, is sized to fill the annulus between the screen and standard casings. In deep water wells, screen strings may be suspended from the intermediate casing by oil-well hangers. The screen is lowered to the desired depth and then the hanger mechanism is operated. The hanger presses against the casing with an action that is made more secure by the weight of the suspended screen. The advantage of the sealed annulus screens is that aquifer material cannot enter the well, but their disadvantage is that— without special ports—gravel pack or formation stabilizer, cannot be installed.

4.3   Auger drilling

Auger rigs vary from small-diameter manual augers for soil sampling to large truck or crane-mounted augers used for drilling shafts that are more than a metre in diameter.

The commonest auger design is the screw auger, with a blade welded in a spiral to a central solid shaft. In small soil augers, the spiral may extend for the first few centimetres of the tip, but with continuous flight augers the auger is supplied in sections, usually 1 m long, with the spiral blade extending the full length of each section.

An alternative design of auger is the 'bucket' auger, which cuts the soil with two blades on the base and then passes the soil up into the bucket. When the bucket is full, it has to be withdrawn for emptying and, as drilling proceeds, sections have to be added to the auger stem. The diameter of mechanical bucket augers is limited by their power source and transmission, but large bucket augers are not common in the UK.

The continuous flight auger is the usual auger used for drilling boreholes. A small exploratory rig with a 75 mm diameter auger flight is shown in Fig. 4.12. The formation being drilled is held on the blades of the auger. Drilling progresses by screwing the auger into the ground for one auger section, and then withdrawing the auger. The auger is then lowered to the bottom of the hole and another section added to the flight.

The samples recovered by ordinary augering are disturbed, but undisturbed core samples can be obtained by hollow-stem augering. The mechanics of drilling are identical, but with the hollow-stem auger the spiral blade is welded to a tube 100 or 150 mm in diameter.

As augering progresses, a core of sediment is forced into the tube or hollow stem, to be recovered when the augers are withdrawn, or by wire-line if a wire-line core barrel is incorporated in the design (Chapter 5).

Auger drilling is a valuable method for rapid formation sampling at shallow depths, or even for drilling small observation boreholes provided

Fig. 4.12 A small exploration rig with augers.

Engineers, 1979; Watt & Wood, 1977).

The excavation in soft formations is usually by shovel, pick and hoe, and the debris is removed from the hole by hoist or conveyor-belt. In hard rock the rock has to be broken up before it can be extracted and this may require a chisel, jack-hammer or even explosives.

The diameter of the hole must be large enough for one, or possibly two, men to work down the hole, but should be as small as possible to minimise the amount of debris to be excavated. A 33% increase in diameter from 1.5 to 2.0 m will result in a 78% increase in the volume of rock to be removed.

A well has to be completed below the water table to develop a storage sump. This is done by keeping the well dewatered—using a surface-mounted, centrifugal drainage pump with its suction hose in a sump in the bottom of the well.

The safety aspects of well digging are commonly overlooked because of the low technology used in the construction. Factors which should always be considered are:

that the formation is soft and cohesive. Augering is not possible in hard rocks, dry sand or in gravels; below the water table, penetration by auger may be impossible because of formation collapse.

4.4 Manual construction

Manual methods are used in the construction of open wells and shafts (Section 2.2). The major factors to be considered in such excavations are methods to maximize the penetration rate and to minimize the danger to the excavators (DHV Consulting

1. The sides of the excavation must be shored up with strong timbers or jacks to prevent collapse.
2. When using pneumatic drills, masks should be worn against the dust.
3. When working below the water table, either a back-up drainage pump or adequate escape ladders should be provided in case of pump failure.

61

Table 4.4 Comparison of construction methods

	Advantages	Disadvantages
Percussion drilling	Low technology rigs, and therefore cheap mobilization and operation	Can drill only to shallow depths because of temporary casing
	Needs small work area	Relatively slow drilling
	Uses little water	
Rotary drilling		
Direct circulation	No limit to depth of drilling	High-technology rig so expensive mobilization and operation
	Fast drilling	
	Needs no temporary casing	May need a large working area for rig and mud pits
		Can use a lot of water
		Mud-cake build-up can make development difficult
DTH and air-flush rotary	No pollution of aquifer by drilling fluid	DTH should not be used in soft, unstable formations
	Needs no water	
	DTH is very fast in hard formations	Drilling depth below the water table is limited by hydraulic pressure
Reverse circulation	Leaves no mud cake	May use great volumes of water
	Rapid drilling in coarse unconsolidated aquifers at large diameters	
Manual construction	Uses low technology, and therefore is cheap where labour is cheap	Restricted to shallow depths

4. When using explosives all precautions demanded for their use in quarries or mines should be followed.

Table 4.4 Compares the various methods of well construction.

5

Formation sampling

The materials recovered from boreholes include samples of the rocks through which the holes are being drilled and the groundwater held in those formations or, in the case of water wells, the water being abstracted from the formations. Formation samples are needed to establish the lithological succession at a site and to assess the hydrogeological characters of the aquifers.

The establishment of a lithological section is a particularly important part of exploratory drilling, but the sampling methods discussed here are also used in observation boreholes and water wells. The most common methods produce disturbed samples, which can be used to identify the formation and indicate the grain-size distribution in aquifer material. They are not suitable for measuring aquifer characteristics and when samples are needed for laboratory tests of porosity and permeability, or even for indisputable formation identification—then undisturbed samples must be obtained.

5.1 Disturbed formation sampling

5.1.1 Augering

Auger drilling can be an excellent method for sampling shallow, unconsolidated formations. Samples from auger drilling are recovered as fragments from the auger flights (Fig. 4.12). The samples recovered are of the true lithology of the formation and sample depth control is good. The sampling technique involves lowering the auger to the bottom of the hole and then spinning it at high rotation to clean debris out of the hole. The depth of the auger tip is recorded, and then the auger is advanced at a slow rotation speed for a fixed depth interval. Rotation is stopped and the augers retrieved by a straight pull. The sample adhering to the auger flight comes from the fixed interval.

The weakness of augering lies in the great difficulty in preventing cross-contamination. As the sample is pulled from the hole it is contaminated by the material above the sample interval. The method cannot be used for taking samples where absolute integrity of the sample is necessary: in trace pollution studies for example. However, if the samples are taken off the auger with care and surfaces of individual fragments cleaned, they can be used for normal pollution studies. Such sampling has been used in England for taking samples of chalk for

64

porewater analysis for nitrate in order to establish vertical nitrate profiles in the aquifer.

5.1.2 Percussion drilling

Chiselling of hard rocks in the unsaturated zone produces a dry sand of crushed rock, which may bear little resemblance to the original rock, although large chips of rock may be used to identify the rock type. The samples are removed by bailer and their identification will follow the same procedure as for samples recovered from drilling fluids in rotary drilling (Chapter 7).

The destruction of the rock fabric and the addition of drilling water means that bailed cuttings can only be used for broad lithological definitions; features such as bedding, porosity or texture cannot be identified. The depth control of the sampling can be good because the hole is bailed clean after each drilling period, but some contamination can still arise from caving formations above.

Once the water table has been reached, water no longer has to be added. The water in the borehole is mixed but, as the drilling progresses, and the drilling water is progressively removed, the water remaining becomes close to true groundwater. Bailed samples, therefore, can be used to give an indication of groundwater quality during exploration drilling when pump sampling may not be possible.

Sampling in unconsolidated soft formations is done by removing the formation by a shell from ahead of advancing temporary casing (Fig. 5.1). The sample arrives at the surface as a slurry, as in hard rock drilling, but in this case the sample is merely a disaggregated sediment. In competent clays or silts a clay-cutter may be used and then reasonable samples can be recovered in the quadrants between the cutter's fins.

Percussion samples, because they are taken from ahead of the casing which prevents wall collapse or caving, generally have good depth control and, despite being disturbed, are representative of the formation lithology and porewater quality. Below the water table, however, porewater quality control is lost because of mixing through the vertical profile of the borehole and because, in unstable sands, flowing sand may fill the borehole from below, so contaminating the formation samples. Despite these potential problems, percussion drilling is widely used for basic sampling surveys.

The bailer samples of disaggregated formations present some problems in sample treatment. All arenaceous sediments contain a certain proportion of clay and silt, which can have a significant effect on the hydrogeological character of the sediments; for example, a few per cent of clay in a sand can reduce its hydraulic conductivity by an order or more. The action of taking a sample in a water slurry can wash out much of the clay or silt fraction from the sand, so care must be taken to avoid underestimating the 'fines' in a sample. A representative sample of the whole contents of a bailer must be taken and the sample should not be

Fig. 5.1 Sampling by percussion drilling.

Shell   Double U 100   Temporary   U 100
         sampler assembly   casing with   sampler liner
                             drive head

washed before it arrives at a laboratory for description (Section 7.2).

5.1.3 Rotary drilling

The drilling-returns from direct circulation rotary drilling (Section 4.2) are a slurry of rock fragments suspended in drilling fluid. The fragments may be broken chips of hard rock or disaggregated sediments, and may range from clays to chips several millimetres across. Examination of the drilling returns and their use in producing a lithological log of the borehole, therefore, can be a complex task. Mud control and mud logging is now an advanced technology in the oil industry, but much less so in the water industry. Nevertheless the compilation of a lithological log requires certain basic measures in either industry.

Formation samples, after being cut by the drill bit, take a certain time to travel to the surface, thus, during continuous drilling a sample is derived from a level above that where the bit is sited when the sample is collected. A correction has to be made to obtain a true sample depth. This depends on the velocity of the drilling mud up the hole, which can be calculated—assuming no circulation losses—from the size of the hole and the drill stem and the drilling fluid pumping rate.

The cuttings in the drilling returns will be contaminated by the drilling fluid as they are being cut but, during their travel to the surface they can be contaminated also with debris eroded from the borehole walls in overlying formations. This debris can represent a large proportion of the cuttings in any mud sample, so a lithological log based on cuttings has to be recorded as a percentile log (Section 7.2). The depth at which any particular lithology is first detected in the cuttings is taken as the depth of the top of the formation that the lithology represents.

The drilling returns, when mud, water or foam is used, are directed to mud pits designed to allow the solid cuttings to settle out of the fluid before it is returned down the borehole (Section 4.2). When mud control is lost and the drilling viscosity is too high, some cuttings will be returned to the borehole in the fluid circulation. These cuttings can severely contaminate the new cuttings and give an erroneous lithological log if one is not aware of the problem.

Samples of cuttings are collected by passing the drilling returns over a container, such as a bucket or drum, set in the mud channel. This container acts as a small mud pit into which cuttings settle, and, in continuous drilling, it is emptied every time the drilling has progressed for a fixed interval. A preferable system of sampling is to have discontinuous drilling, where drilling stops at the top of the sampling interval but fluid circulation continues until no further cuttings are being returned. Drilling then continues over the sampling interval and the cuttings are collected. The depth control of the cuttings taken in this way is improved and cross-contamination of samples is reduced.

The samples recovered from the sampling bucket may contain drilling mud and the addition of a solution of polyphosphate, such as Calgon, will help to disperse this mud and clean the sample. However careful samples from direct circulation drilling, where a liquid drilling medium has been used, are treated, loss of finer fractions of the formation is almost inevitable, and any interpretation based on the samples must take this into account.

Rotary drilling using air flush (Section 4.2.1) avoids some of the problems encountered when using drilling fluids. The air flush is not returned to the borehole, and therefore there is no cross-contamination by recirculated cuttings. While drilling through the unsaturated zone, the cuttings are not mixed with mud or water and therefore do not need cleaning before inspection. Air-flush sampling, however, cannot avoid contamination of samples by caving formations, and drilling below the water table will produce a slurry of cuttings in native groundwater. When using air-flush drilling for investigating sites contaminated by hydrocarbons, oil traps must be used on the air line to avoid contamination of the samples and *in situ* formations by compressor oils. It must also be recognized that in such a situation air flush will strip any volatile pollutants from the formation samples.

The samples derived from reverse circulation drilling are returned to the surface up the drill stem and do not come into contact with overlying formations. The samples, however, may be contaminated by caved debris carried to the drill bit in water flowing down the annulus around the drill stem. The velocity of the return flow with reverse circulation is high, so that the larger fragments of the formation can be returned to the surface. This is a great advantage when drilling and sampling gravel formations, because the coarser fractions can be recovered uncrushed and so sample identification is much easier. A problem with sampling from reverse circulation rigs is caused by the high velocity of the returns, which emerge as a jet. The samples can be collected in a sieve or a bucket held in the jet or by diverting part of the jet. In any case, the loss of the finer fractions of a sediment is inevitable.

5.1.4 *Storage of disturbed formation samples*

The storage and clear labelling of formation samples is extremely important to avoid future confusion. With disturbed material, samples of about 500 g in weight should be taken on removal from the borehole, and should be stored in sealable plastic containers, either bags or jars. Each sample must be labelled with waterproof ink, either directly or, preferably, using a separate label. The label should contain all relevant information and be clearly visible when the sample is picked up. A typical sample label is shown on Fig. 5.2. The label should be durable, preferably plastic, to avoid being spoiled by the wet sample. It is recommended that samples should be double bagged, with the label put between the two bags to keep it clean, clear and secure.

Fig. 5.2 Sample label.

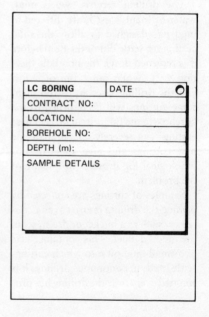

5.2 Undisturbed formation sampling

The action of drilling to obtain any sample will disturb that sample to some extent; 'undisturbed formation sample', therefore, is a relative term. Even in a stick of apparently unbroken core the drilling will have heated the core and the drilling fluid will have penetrated the periphery of the core. The sampling methods described below solve the problem of obtaining truly undisturbed samples with various degrees of success, depending on the formation being drilled and whether or not drilling fluids have been used.

5.2.1 Hollow-stem augers

The central part of a hollow-stem auger, as the name implies, is a hollow tube, and as the auger advances a core of the formation is forced inside the auger stem. When the augers are withdrawn, the core can be extruded intact. This method of sampling can only be used to investigate unconsolidated formations capable of being augered, but it can be used to obtain samples from above and below the water table.

5.2.2 Percussion drilling

The most commonly used equipment for obtaining undisturbed samples by percussion drilling is the 100 mm ID open-drive sampler or the U100 sampler (U4 in Imperial Units). A U100 tube assembly (Fig. 5.3) com-

Fig. 5.3 U100 sampler.

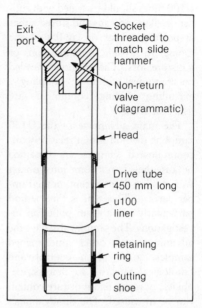

prises a 450 mm-long steel tube of 100 mm ID to which a cutting shoe is screwed. The steel tube and shoe can be machined so that an aluminium liner can be inserted in the tube. The tube is then screwed to a U100 head, which in turn is attached to a slide hammer below a sinker bar.

The borehole to be sampled is cleaned out to the bottom with a bailer, and then the U100 assembly is lowered slowly to the bottom. Drilling then proceeds by blows of the slide hammer until 450 mm penetration has been made—the blows are counted to give an indication of the formation hardness. The U100 assembly is lifted to the surface and the U100 tube is unscrewed. Either the liner or the tube itself is kept until the sample can be

69

removed for examination. The full U100 tubes should be sealed with rubber or alloy end-caps and stored either in polythene bags or lay-flat plastic tubing to avoid contamination or moisture loss. The materials needed on-site to deal with U100 sampling—excluding drilling equipment—are listed on Table 5.1.

The main advantage of the U100 sampling method is that relatively uncontaminated samples are obtained because they do not come into contact with drilling fluids except on their upper surface. This makes the method particularly valuable in pollution investigations. The samples also have the advantages of other undisturbed samples: good depth control and lithological integrity so that permeability, porosity and moisture content can be measured. The disadvantages of U100 sampling are that it cannot be used in consolidated rocks or sediments containing boulders, because of damage to cutting shoes and tubes; sample recovery in soft clays or flowing sands may be very poor. In soft formations, sample

recovery can be improved by incorporating spring clips or core lifters in the cutting shoe. U100 assemblies are easily damaged, so a good stock of spares should be carried:

The sample is extruded from the tube or liner by a manual or electric hydraulic ram (Fig. 5.4). The sample appears as a core, but very commonly it is in three parts, the lower part (close to the cutting shoe) is virtually undisturbed, the middle part is fractured and the upper 100 mm or so have been disturbed by the percussion technique. If the whole core is needed in an almost undisturbed state, then a double U100 can be used in which two U100 tubes are connected with a collar. Sampling is as for a normal U100 but the sampler is driven in up to 900 mm. The formation held in the cutting shoe and in the top tube is discarded and only that held in the bottom tube is kept for examination.

Formation sampling in pollution investigations may require special precautions. When sampling for material polluted by organic materials

Table 5.1   Check-list for U100 sampling site

---

1. U100 hydraulic extruder.
2. Rolls of 150 mm diameter lay-flat plastic tube.
3. Heat sealer for lay-flat tube.
4. Supply of U100 tubes.
5. Supply of U100 liners.
6. Supply of end-caps—2 per U100 tube.
7. Plastic labels and waterproof pen.
8. Notebook
9. Cold boxes or freezer for temporary core storage.

Fig. 5.4 U100 core extruder on-site.

Fig. 5.5 Through-flow sampler.

no organic material such as grease, which could introduce contamination, must be used on the U100 assembly. The assembly must be degreased before sampling begins and the samples must be sealed in airtight bags to avoid the loss of volatiles. In obtaining samples for microbiological assays, even greater care must be taken to avoid sample contamination. The U100 tubes, liners and end-caps must be sterilized and kept in sterile bags before use. The entire U100 assembly has to be flame sterilized before it is lowered to take each sample and, on retrieval, sample handling should be as careful as possible. The reader should seek specialist advice on degreasing and sterilization techniques.

Samples of unconsolidated sediments or soft formations such as chalk can be obtained by a through-flow sampler. This type of sampling is restricted to shallow depths but it is ideal for rapid exploration of a restricted site. The sampler is driven by a hydraulically operated jack-hammer (Fig. 5.5) to the pre-selected depth. The formation passes up the hollow sampler and out of side vents as the sampler penetrates, so that the sample is retained when the sampler is withdrawn.

The sampler can be machined to incorporate a split core-barrel or a plastic liner. Recovery from fine, soft formations is good, but pebbles prevent sampling and soft clays with low shear strengths are difficult because the sampler just pushes through the sediment. The sampling depth control is good. The samples taken from the split core barrel provide a good lithological log and can be used to get samples of porewater for pollution profiling.

Various sampling devices similar to

71

the U100 or through-flow samplers have been developed for sampling soft formations from shallow boreholes. These samplers are more commonly used in site investigations for civil engineering works than in groundwater resource or pollution investigations. They include:

Thin-walled stationary piston sampler used for soft non-cohesive sediments (shear strength < 50 kN/m^2. The piston sampler is forced into the formation by a steady force from a jacking system, rather than by blows as in a U100 sampler (BS 5930 p. 24).

Standard penetration test (SPT) or Raymond sampler. The end of an SPT sampler has a cutting shoe and is hollow, rather like a through-flow sampler in that a formation sample is forced into the tip and can be recovered. The sample is small, generally fractured, and, by the time that it is removed from the sampler, more akin to a disturbed sample. An SPT sample, however, can be valuable for providing lithological control for a log (BS 5930, 1981, p. 33).

Compressed-air sand sampler or Bishop sampler; designed to obtain samples of non-cohesive sands below the water table, a situation where U100 sampling may fail (BS 5930, 1981, p. 26).

Continuous drive sampler or Delft sampler; specially designed for sampling the thick soil sequences of Holland, suitable for very soft, non-cohesive sediments. The sampler is pressed into the ground by continuous pressure and the core passes back into a plastic liner. One-metre extension tubes are added as the sampler progresses, allowing continuous sampling. A depth of 18 m is common, but greater depths can be reached with heavier equipment.

5.2.3 Rotary Drilling

The recovery of undisturbed formation samples by rotary drilling is by coring using direct circulation (Section 4.2.1). Equipment available for coring is versatile and variable because of its importance to the oil and mineral exploration industries. Coring is carried out by means of a special drill-bit assembly known as the core barrel (Fig. 5.6). The core barrel is a hollow steel tube generally made up of four parts: head, tube, reaming shell and core bit. The head is threaded to be compatible with the drill stem. The tube and reamer are both hollow tubes but the reamer has raised, diamond-studded plates on its face to clean and ream out the well face behind the bit. The core bit is a steel tube with the bottom rim studded with diamonds or tungsten carbide inserts. Between the bit and the reamer is an internal ring of steel ribs called the core lifter which retains the core in the barrel when it is being lifted from the hole.

In hard rock formations a single-tube core barrel can be used (Fig. 5.6). The rotation of the drill stem and the circle of the bit rim cuts a circular plug or core of rock, which passes up into the core barrel. The drilling fluid passes down the drill stem, between the core and the barrel, to cool the bit. This single-tube core barrel is suitable only for massive, strong, uniform rock because the drilling fluid will flush out all but the most cohesive cores.

Fig. 5.6 Core barrel types.

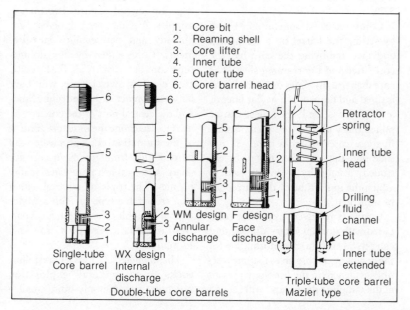

1. Core bit
2. Reaming shell
3. Core lifter
4. Inner tube
5. Outer tube
6. Core barrel head

Single-tube Core barrel

WX design Internal discharge

WM design Annular discharge

F design Face discharge

Double-tube core barrels

Retractor spring

Inner tube head

Drilling fluid channel

Bit

Inner tube extended

Triple-tube core barrel Mazier type

The most commonly used core barrels are double-tube types, of which several designs are available: rigid, swivel (WX and WM designs) and face ejection swivel type (F design) (Fig. 5.6). These core barrels each have an outer and an inner tube, and the drilling fluid passes between the tubes and so does not wash away the core. In the rigid and swivel types the drilling fluid does wash over the bottom few centimetres of the core, and in soft formations the core may be lost. In the face ejection swivel type the drilling fluid is ejected at the rim of the bit and does not touch the core, so that this barrel is the most suitable for soft or friable formations.

In a triple-tube barrel a liner, usually split, is incorporated in the inner tube and the core is extracted inside the liner. A rectractable triple-tube barrel (Triefus or Mazier type) has been developed for extremely soft, friable sediments. This design has a spring-loaded inner tube and liner. When drilling in soft material the spring holds the leading edge of the inner tube ahead of the core bit, so that the core is totally protected from the flushing medium.

An alternative to the triple-tube assembly is the use of Mylar plastic liner. This material is a durable, flexible, translucent plastic supplied in rolls up to 200 m long and in widths suitable for most core barrel sizes. The Mylar is rolled into a cylinder, inserted into the inner tube, trimmed to size then clamped into position by a retain-

73

ing ring set behind the core lifter (Fig. 5.7).

Core is retrieved from the core barrel by setting the barrel on a clean rack and then removing the core bit, the core lifter and the reaming shell. The core then usually slides out if the barrel is tilted and tapped. If a Mylar liner or a split liner is used, then these can be pulled out with the core intact inside. In those cases where a core will not slide out, then the core has to be extruded, usually by hydraulic ram to which the inner tube of the barrel can be attached.

Coring is relatively expensive compared with normal rotary drilling, but can provide invaluable information for the geologist. A core gives incontravertible evidence on the lithology present at the sampling depth, and will act as a base for comparison with geophysical logs and logs of disturbed formation samples. A good core can be used for porosity and permeability measurements. Cores taken by any double-tube core barrel with air flush, or by face ejection swivel type will have minimal contact with the drilling fluid, and the central part of the core can be used to determine the moisture content of the formation. The porewater can be extracted from the clean core for analysis to establish porewater quality profiles. The triple-tube barrel, either as a retractable type or with a Mylar liner in a double tube-barrel, gives particular clean samples in soft formations.

The diameter of cores of crystalline rocks from mineral exploration boreholes is commonly quite small—

Fig. 5.7 Inserting Mylar liner into core barrel.

Mylar liner   Core barrel

about 19–55 mm (R–N core barrel sizes, Appendix A.II.1), because these cores are taken principally for lithological identification, and any invasion by drilling fluid is of minor importance. Cores of sedimentary formations taken during groundwater investigations generally should be chosen with large diameters, usually between 76 and 113 mm (core barrel sizes H, P or S). This larger diameter will give sufficient core material for laboratory tests on grain size, porosity and permeability. There will be some mud infiltration into the edge of the core but, with a 100 mm diameter, there is a good probability that the centre of the core will be genuinely undisturbed. If core is used to obtain porewaters for analysis, then only the centre of the core should be taken to ensure, as far as possible, that contamination by drilling fluid is avoided.

5.2.4 Storage of undisturbed samples

The cores, on retrieval, are placed in a core box for transport and storage. The box is divided into longitudinal sections wide enough to hold the core fairly firmly (Fig. 5.8). The boxes should not be too large to be lifted when full by two men. The core sections have to be labelled top and bottom with relevant depths, and any sections not cored should be indicated by spacer blocks with their depths.

The boxes must be clean and, in cases where the moisture content is to be measured or the porewater extracted, the cores must be sealed in sleeves of lay-flat polythene to avoid evaporation. Cores of unconsolidated sediments should be extruded and kept in sleeves to keep them intact—stockingette can be used, but Mylar is ideal. Samples which are going to have their porewaters analysed for nitrate or organic determinands, must be chilled immediately or frozen to avoid biodegradation of the determinands. Samples for microbiological examination similarly must be kept in sterile sleeves of polythene lay-flat and stored in the dark at 4°C to avoid bacterial or fungal growth.

Fig. 5.8 Core box layout.

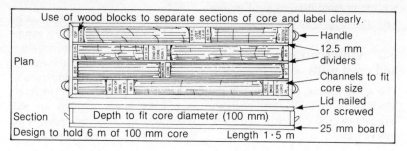

75

6

Groundwater sampling

6.1 Production boreholes

Sampling groundwater from existing production boreholes has the advantage that the sampling facilities are usually good and the water will be fresh groundwater with minimal effects from well construction or from a static water column in the borehole. Many production wells, however, have chlorination facilities and care must be taken to take a sample from a raw-water tap at the well head as the chlorine will affect both the chemistry and bacteriology of the water. The water sample is a composite sample of the groundwater in the aquifer, although geophysical logging can be used to indicate inflow zones and so the approximate source within the borehole. The composite nature of these samples means that they can be used to study regional variations in groundwater quality, but they are of less use in more detailed local studies such as pollution investigations.

6.2 Observation boreholes and open wells

The water column in an observation borehole is open to the atmosphere so its quality will be affected by a complex mix of factors:

1. Part of the column may be static and stale.
2. Dissolved gases will exsolve, affecting the pH and carbonate equilibrium.
3. Atmospheric oxygen will enter the system, affecting the iron equilibrium.
4. Light may allow organisms to grow, so affecting the organic content of the water and many biologically-mediated reactions.
5. Foreign material (solid or liquid) may fall into the column, polluting the water.
6. The well construction materials may react with, and contaminate, the water.

In addition, a borehole screened in several aquifers, or indeed in one thick aquifer, will intercept water with different heads, leading to flow up or down the borehole and so a mixed water.

It is recommended that no sample be taken from a borehole without pumping, to ensure that fresh groundwater has totally replaced the static column and that the sample will be representative of the groundwater. However, in practice, many samples are still taken without prior pumping. Two main types of samplers are used.

6.2.1 Casella-type sampler

This is a cylinder, bottle or bucket lowered into the groundwater to take a 'grab sample'. The sample quality is improved if the sampler is used to bail out the water column to remove the stale water first.

6.2.2 Depth sampler

Several designs of depth samplers are available but most incorporate a cylinder which can be opened and closed remotely from the surface. The sampler is lowered to a pre-selected depth, filled with water and then sealed enabling the sample to be brought to the surface. Sequential sampling from the water surface to the bottom of the borehole enables a profile of the water column to be measured. Those samples taken from adjacent to screens or open aquifer may be close to native groundwater, but where there is flow up or down the borehole, mixing is inevitable, so that no profile is likely to accurately reflect the true water-quality profile in the aquifer.

The main problems associated with Casella and depth samplers can be overcome by using a pump to obtain samples. Where the water level is within about 8 m of the surface, samples can be obtained by surface-mounted pumps. Deeper water tables will need a submersible pump. A convenient portable sampling set comprises a 100-mm diameter submersible pump, flexible plastic-fabric rising main and a generator. Forty-five mm diameter submersible pumps, either

using conventional impellers or pneumatic bladder or piston mechanisms, are available for sampling small-diameter boreholes. A submersible bilge pump operated by a car battery can be used for sampling such holes with a shallow water level.

A borehole should be pumped long enough to remove four times the volume of the water column, to ensure that reasonably representative groundwater is being sampled. The sample will still suffer from the drawback of being a composite sample of water from the full aquifer thickness. Multiple borehole completion (p.24) can overcome this problem in part but, with an open borehole, the only way to ensure that samples taken are true groundwater from specific horizons is by straddle packer sampling.

6.2.3 Straddle packer sampler

The principle of a straddle packer is quite simple, but the weight of a packer assembly and ancillary equipment for boreholes of 100-mm diameter or more, means that a drilling rig or tripod and winch is needed to use it efficiently. The packer assembly comprises an axial pipe on which are mounted two inflatable rings at a fixed spacing. The rings may be inflated by air line or water lines from the surface (Fig. 6.1) or, less commonly, by screwed compression. The inflation has to be sufficient for the assembly to be lowered easily down the hole when deflated, but to seal the borehole when inflated. The material of the ring must

Fig. 6.1 Straddle packer assembly.

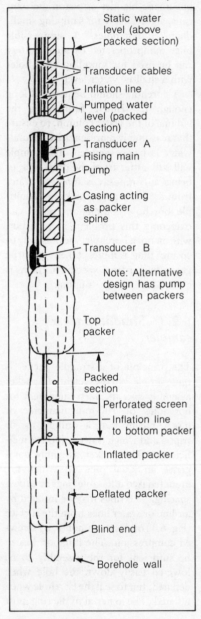

Static water level (above packed section)

Transducer cables

Inflation line

Pumped water level (packed section)

Transducer A

Rising main

Pump

Casing acting as packer spine

Transducer B

Note: Alternative design has pump between packers

Top packer

Packed section

Perforated screen

Inflation line to bottom packer

Inflated packer

Deflated packer

Blind end

Borehole wall

be robust to withstand harsh treatment. The spacing of the rings may be between one and two metres—the longer the spacing the more unwieldy is the assembly. The sampling pump can be mounted between the packers or above the packers. The latter configuration is easier to assemble, but can only be used when the pump itself is below the pumping water level in the borehole. The whole assembly is mounted on the end of a rising main, up which the water is pumped.

The straddle packer can only be used effectively in open holes in reasonably sound rock, because the action of inflating and deflating the packers can cause unstable rocks to cave and jam the assembly down the borehole. Also, boreholes in unstable rocks may have such variable cross-sections due to fissuring or caving that sealing with the packers becomes very uncertain.

The packer assembly is lowered to the first pre-selected sampling points with its packers deflated. The packers are then inflated to seal off the packed interval. The pump is switched on at as low a rate as possible, then, if the horizon proves productive, the rate is increased. The water level above the pump can be checked by transducer mounted in the packed interval. The seal of the packer can be checked by the non-reaction of another transducer mounted above the packer assembly. Ideally, the seal of the lower packer also should be checked by the non-reaction of a transducer set below the lower packer. The packed interval should be pumped until four times the volume of the interval has been removed or until quality parameters such as conductivity, pH, temperature or Eh have stabilized before a sample is taken.

After the first sample has been taken, the packers should be deflated and the assembly lowered to the next interval down the borehole. To get a continuous profile, the packed intervals should overlap adjacent ones by about a third.

The straddle packer also can be used in a similar manner to perform pumpings tests (or recharge tests) to build up a permeability profile through the aquifer. The changes in water levels or head in these tests are measured by pressure transducers.

Permanent sampling installations have been designed to overcome the problems of atmospheric access and of mixing of different waters in open boreholes. The two main types of system may be described as *in situ* samplers and multiple-port samplers.

6.2.4 In situ *sampler (WRc design)*

A cylinder is divided into two chambers separated by a non-return valve (Fig. 6.2). Water flows into the upper chamber through filtered ports, and then through the valve into the lower chamber. The sampler is connected to the surface by two narrow tubes, the first one passes into the top of the lower chamber, while the second one passes to the bottom of the lower chamber. The groundwater sample is obtained by forcing compressed gas down the first tube, so forcing the

water to the surface through the second one. The two tubes must be clearly labelled by metal or plastic tags at the surface.

A borehole is drilled to its total depth at a minimum diameter of 150 mm and then geophysically logged to define zones of different water quality and to select sampling intervals of interest. The borehole is then backfilled with bentonite or aquifer material to 1 m below the lowest selected sampling depth and a sampler lowered to the selected depth. Gravel is put in the borehole to completely surround, and extend to 1 m above, the sampler. The interval to 1 m below the next sampling depth is then filled with either bentonite pellets or a bentonite

Fig. 6.2 WRc *in-situ* sampler.

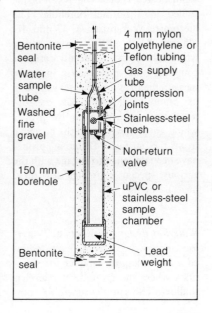

79

slurry to form an impervious seal between the sampling intervals. Samplers are installed in this way to the shallowest sampler, and then the borehole is backfilled to the surface.

The depth samplers, after purging to remove any effects of drilling, will give samples of groundwater unaffected by surface influences or inter-aquifer reactions. The samplers can be made of uPVC for normal use or of stainless steel with Teflon tubes for studies where organic compounds have to be determined. The main problem with such samplers is that once they are installed they cannot be retrieved for maintenance or repair.

6.2.5 Multiple-port samplers

Westbay System: This involves a sophisticated system which can be installed in a normal borehole. The piezometer comprises a 37 mm ID (48 mm OD) uPVC casing string which incorporates specially designed couplings for piezometric pressure measurements, groundwater sampling or pumping. Samples or pressure measurements are taken by probes which are lowered down the borehole from the surface. Different probes have been designed to engage with the various special couplings for these measurements.

Waterloo design: Workers at Waterloo University in Canada have designed a multiple-port sampler for situations where the groundwater level is shallow. The sampler assembly comprises a 50 mm ID well casing perforated at fixed intervals along its length. Each perforation is connected to the surface by a narrow tube which passes along inside the casing and through which a sample can be drawn by vacuum or by peristaltic pump.

6.3 Collection, recording and analysis of groundwater samples

The material dissolved in groundwater is usually in equilibrium with the host rock and the physical conditions at depth in the aquifer. This equilibrium is disturbed when the water is pumped to the surface and the quality of the groundwater can change fundamentally within minutes of leaving the borehole. This change in quality means that the groundwater quality determinands can be divided into four groups on the basis of their sampling and analytical needs:

1. Major cations and anions.
2. Unstable determinands.
3. Minor determinands.
4. Microbiological analyses.

The decision as to which groups of determinand to analyse for will depend on the particular groundwater investigation being undertaken, but in all cases the fieldworker is strongly advised to discuss the whole programme with the chemist who will be doing the analyses. The chemist will be able to advise on sampling procedures, on the size of samples needed and will be able to supply bottles needed for specialized samples.

Fig. 6.3 Groundwater sampling record sheet.

LC BORING	GROUND WATER SAMPLE: REFERENCE No:
DATE	LOCATION
METHOD OF SAMPLING	
SITE SKETCH	NOTES
ON-SITE MEASUREMENTS COLOUR TASTE SMELL TEMPERATURE °C CONDUCTIVITY: μS/cm	Eh pH ALKALINITY DISSOLVED OXYGEN IRON
LABORATORY DETERMINANDS REQUIRED. TICK AS REQUIRED RANGE EXPECTED RANGE EXPECTED	
CALCIUM MAGNESIUM SODIUM POTASSIUM CHLORIDE SULPHATE ALKALINITY (MO & PP)	NITRATE NITRITE AMMONIUM O-PHOSPHATE IRON pH CONDUCTIVITY
OTHER DETERMINANDS	

In the field, the person taking the samples should record on a record sheet (Fig. 6.3) sampling details which should include the method of sampling, the colour, smell, taste and temperature of the water and the condition of the well or borehole (closed or open, clean or dirty). The determinands required should also be listed with, if possible, some indication of concentrations expected and accuracy of analysis required. A copy of this record sheet should be sent with the sample, which should also be clearly labelled to avoid confusion in the laboratory. The label (Fig. 6.4) can be tied or stuck to the sample bottle, but all writing should be with waterproof ink.

Suspended solids in the water samples can seriously affect analytical results, so that all samples should be filtered before being sent for analysis.

Fig. 6.4 Groundwater sample label.

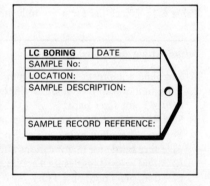

The filtering can be done through a normal filter funnel with a fine (GF/C) filter paper, but would be better done through a 0.45 μm filter paper under pressure (Hitchman, 1983). The equipment needed for pressure filtration would have to be made before the project began.

81

The four groups of determinands are briefly discussed separately below, and a check-list for a groundwater sampling programme is given in Table 6.1.

6.3.1 Major cations and anions

These will commonly make up more than 90% of the minerals dissolved in groundwater. They are used to quantify the gross quality and classify the type of groundwater (Hem, 1985), and comprise the 'standard' groundwater analysis. They include chloride, sulphate, alkalinity (bicarbonate), nitrate, potassium, sodium, magnesium and calcium. If the concentrations of these eight determinands are converted to milli-equivalents per litre (Appendix A.III), then the total cations should balance the total anions within 10% for an acceptable analysis. It is recommended that such ionic balances should be used as routine quality control for the analytical services.

An estimate of the total dissolved solids (TDS) in the water can be made from the electrical conductivity (EC) of the water. The conductivity is measured with a portable EC meter (Fig. 6.5) and the TDS read off a correlation graph such as Fig. 6.6. The relation between TDS and EC is dependent on the proportions of the various major ions in the water, but for any specific water type the relationship is linear.

Table 6.1 Check-list for a groundwater sampling trip

Before setting out, have you got:

1. Sampling equipment—pump, depth sampler or Casella-type sampler.
2. Filter funnel with box of GF/C filter papers or pressure filter unit and box of $0 \cdot 45$ μm filter pads
3. Sample bottles for all necessary groups of determinands.
   For each sampling point you will need:
   a) 1 litre bottle for major cations and anions.
   b) 50 ml primed bottles for Fe, NH_4, NO_3 and NO_2.
   c) 50 ml glass bottle for TOC.
   d) Special bottles and equipment for minor determinands as advised by analysts.
4. Sample labels and record book.
5. Waterproof pen or pencil.
6. Key or spanners to open the boreholes and wells.
7. pH, Eh, EC, DO and temperature meters with operating manuals and standards for calibration.
8. Portable laboratory.

Fig. 6.5 Equipment for water analyses in the field.

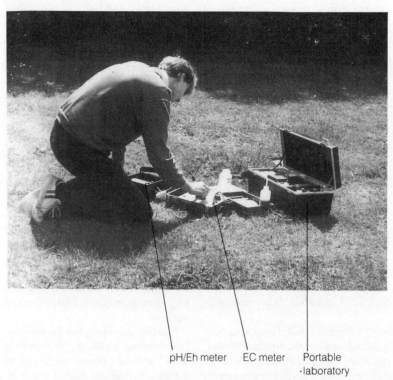

pH/Eh meter   EC meter   Portable
·laboratory

Most of the major ions are stable, but it is advisable for all samples to be analysed within 48 hours. Samples should not be stored in soda glass if sodium is being measured. The suite of eight major cations and anions *can* be determined with about 100 ml of sample, but it is recommended that the sample for this analysis is one litre. The sample should be collected in a clean one-litre glass or plastic bottle with a screw-thread or ground-glass stopper. The sample should fill the bottle to exclude air.

The alkalinity of natural ground-water is largely the result of bicarbonate ions, but the bicarbonate content of water is dependent on the equilibrium:

$$2HCO_3^- \rightleftharpoons CO_2^{2-} + H_2O + CO_2 \uparrow$$

This equilibrium may change considerably as a deep groundwater is brought to the surface and degasses. The equilibrium is also dependent on the water temperature, so for an absolutely true measure of the bicarbonate the analysis should be done at the well head with those of the unstable determinands; this is rarely done.

83

Fig. 6.6 Relationship between dissolved minerals and electrical conductivity of water. After Hem, 1985.

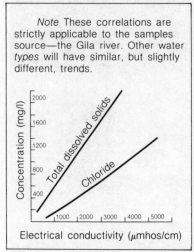

Note These correlations are strictly applicable to the samples source—the Gila river. Other water types will have similar, but slightly different, trends.

6.3.2 Unstable determinands

Several determinands change so quickly on release from the aquifer that they can only be measured effectively at the well head. These include temperature, dissolved gases (CO_2 and O_2), Eh, pH, and alkalinity (bicarbonate).

Meters are available with specific probes to analyse directly for dissolved oxygen and Eh, but care has to be taken to bleed the water sample directly from the well head through a closed chamber holding the probes, so that there is no access for oxygen. The alkalinity and CO_2 can be measured by titration. Meters are also available for the measurement of pH and temperature, but these can be measured in an open container such as a beaker. The various meters should be operated according to the manufacturer's manual.

They are electronic instruments and should be treated and maintained with care, particularly their probes, which are delicate. The meters will need to be calibrated each time they are taken in the field, or regularly during extended field use.

The iron solubility is strongly controlled by pH, Eh and dissolved oxygen. Access of atmospheric oxygen to the groundwater will change the Eh and precipitate most of the dissolved iron within minutes (p.101). Dissolved iron can be measured on-site by colorimetric methods using a portable laboratory (Fig. 6.5), but if this is not possible, then a 50 ml groundwater sample should be taken in a small glass or plastic bottle, primed with nitric acid to make a $0 \cdot 5$ molar solution when the water is added. The acid will keep the iron in solution for laboratory analysis.

Nitrate and nitrite can be removed by microbial action if the sample is stored at room temperature, and so samples for nitrate should either be stored in a fridge or treated with a biocide. It is recommended that 50 ml sample should be taken as for iron, but with the bottle primed with $0 \cdot 5$ ml of mercuric chloride solution of $0 \cdot 4$ g/ litre strength.

Ammonium can be lost from a sample by biodegradation or by evaporation during storage, so it is recommended that a 50 ml sample be taken in a bottle primed with $0 \cdot 25$ ml of 3 molar sulphuric acid to fix the ammonium.

Total dissolved organic carbon (TOC) is used as an index of organic contamination of groundwater. Samples for TOC analysis should be taken in 50 ml ground-glass stoppered bottles previously acid-cleaned to remove any grease or organic material.

The primed bottles for iron, nitrate + nitrate and ammonium should be prepared by a chemist before being taken into the field for the sampling. The bottles should have either plastic screw caps or ground-glass stoppers, and should be clearly labelled to show for which determinand they are intended. The TOC bottles need to be cleaned in a laboratory.

6.3.3 Minor determinands

These include trace metals, radio-isotopes and trace organic compounds. The sampling in all these cases is extremely important and advice from specialists must be obtained.

In general, the sample containers should be pre-cleaned in the laboratory, and great care has to be taken to avoid cross-contamination during sampling and transport. When sampling for tritium (^3H) the bottle should be rinsed repeatedly in the water to be sampled, and care has to be taken to avoid touching the sample; sweat contains high levels of tritium. The sampling equipment for trace metals should be chosen so as to avoid equipment made of the metals being measured. Similarly, the containers for trace organic samples must have been scrupulously cleaned in the laboratory, and no plastic, grease, skin or fumes must touch the sample. Measurable cross-contamination is difficult to avoid because the analytical methods can measure such low levels—nanograms per litre in the case of trace organics. One nanogram per litre represents about one teaspoon in $5000m^3$, or one million Imperial gallons.

6.3.4 Microbiological analyses

These have to be done by specialist laboratories. Samples of water are taken in sterile bottles supplied by the laboratory, taking great care to avoid any possible contamination of the sample. The bottle should be open for the minimum time needed to take the sample, and the sample, the inside of the bottle and the bottle stopper should not be touched during sampling. Bacterial or viral populations can change very quickly, so the sample must be kept at about 4°C by packing in ice, and should be delivered to the laboratory on the day of sampling.

Portable analytical kits are available for assaying pathogenic bacteria in remote areas where samples cannot be taken to a laboratory.

7

Logging of boreholes and wells

A borehole log is either a verbal or pictorial description of some aspect of a borehole; its construction, the lithological sequence that has been drilled, the occurrence of groundwater or the geophysical characteristics of the borehole. The examination of either formation or groundwater samples should not be done without the help of the driller's or geophysical log of the borehole. The original driller's log, particularly his record of the rate of penetration, can also be used to characterize the formations drilled. Geophysical logs can be useful in identifying the source of samples and vice versa.

7.1 Driller's logs

The daily records of drilling made by the driller (Tables 4.2 and 4.3) are the basis of the driller's log. The information in the records has to be abstracted to build up a complete picture, as seen by the driller, of the formations drilled, the groundwater encountered and the construction of the borehole or well. The information should be recorded as a pictorial log, as shown, in part, on Figure 7.1.

7.2 Lithological logs

Sample descriptions are best done in a laboratory with full facilities, but where this is not possible an adequate examination can be done on-site, provided that a clean bench with a water supply and a light source is available. The examination can be done by hand lens, but it is best to use a binocular microscope. Laboratory facilities are essential for porosity or permeability determinations, but grain-size analyses can be done on-site if this is absolutely necessary.

Undisturbed core samples are logged to identify those samples required for analysis in the laboratory, and to make a preliminary assessment of the lithological succession. This logging should be done, if possible, on-site—as the core is recovered from the borehole. The log should be written to include a lithological description, stratigraphic identification and a judgement as to the need for further analysis (Fig. 7.2).

Core samples can be analysed in the laboratory, using standard petrographic techniques. These sample analyses, which should include petrographic descriptions, grain-size

Fig. 7.1 A driller's log.

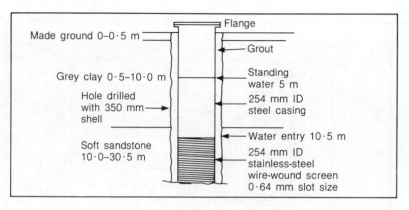

Flange
Made ground 0–0·5 m
Grout
Grey clay 0·5–10·0 m
Standing water 5 m
Hole drilled with 350 mm shell
254 mm ID steel casing
Soft sandstone 10·0–30·5 m
Water entry 10·5 m
254 mm ID stainless-steel wire-wound screen 0·64 mm slot size

analyses, and porosity and permeability measurements, will give definitive data, against which geophysical logs and disturbed formation samples will be compared to establish a comprehensive lithological log.

Disturbed formation samples are examined in sequence and their descriptions compared with all other available data to gradually build up the log. Each sample should be carefully washed in a Petrie dish, to remove the drilling additives but not the formation fines. The washed sample will represent the formation drilled, together with contaminating fragments from above. The sample should be recorded as a percentile log, then each fraction described in greater detail to build up a fuller picture of the formation (Fig. 7.3). Sedimentological texts are available which describe the laboratory techniques and provide charts to assist in estimating sample percentiles, sorting and roundness (Tucker, 1982),

Fig. 7.2 A core log.

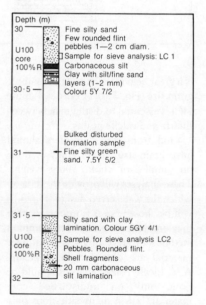

Depth (m)
30
U100 core 100%R
Fine silty sand
Few rounded flint pebbles 1–2 cm diam.
Sample for sieve analysis: LC 1
Carbonaceous silt
Clay with silt/fine sand layers (1–2 mm)
30·5
Colour 5Y 7/2

Bulked disturbed formation sample
31
Fine silty green sand. 7.5Y 5/2

31·5
Silty sand with clay lamination. Colour 5GY 4/1
U100 core 100%R
Sample for sieve analysis LC2
Pebbles. Rounded flint.
Shell fragments
20 mm carbonaceous silt lamination
32

and a selection of such charts is given in Appendix A.V. A sand ruler, comprising a graded set of sieved sand samples,

Fig. 7.3 A lithological log.

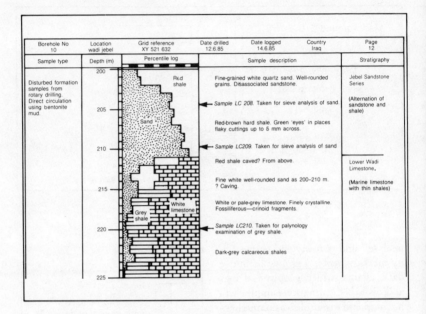

Borehole No 10	Location wadi jebel	Grid reference XY 521 632	Date drilled 12.6.85	Date logged 14.6.85	Country Iraq	Page 12
Sample type	Depth (m)	Percentile log	Sample description			Stratigraphy

| Disturbed formation samples from rotary drilling. Direct circulation using bentonite mud. | | Red shale / Sand / Grey shale / White limestone | Fine-grained white quartz sand. Well-rounded grains. Disassociated sandstone.

← *Sample LC 208*. Taken for sieve analysis of sand.

Red-brown hard shale. Green 'eyes' in places flaky cuttings up to 5 mm across.

← *Sample LC209*. Taken for sieve analysis of sand

Red shale caved? From above.

Fine white well-rounded sand as 200–210 m. ? Caving.

White or pale-grey limestone. Finely crystalline. Fossiliferous—crinoid fragments.

← *Sample LC210*. Taken for palynology examination of grey shale.

Dark-grey calcareous shales | | | Jebel Sandstone Series

(Alternation of sandstone and shale)

Lower Wadi Limestone.

(Marine limestone with thin shales) |

is a great help in estimating the sample grain size (Fig. 7.4). A bottle of dilute HCl is essential to distinguish between quartz and calcite grains.

Sand from target aquifers should have grain-size (sieve) analyses done on samples of about 500 g weight. These analyses will provide the data on which the well screen design (Fig. 3.3) will be chosen, so it is important that they are done correctly. The common methods of sieve analysis of aquifer material are discussed in Appendix A.V. Ideally, sieve analysis should be done only on undisturbed core material, but in many situations only disturbed formation material is available. In that case, the sample for analysis has to be chosen from the

Fig. 7.4 A sand ruler.

least-contaminated material, as judged from the percentile log (Fig. 7.3).

The examination of consolidated formations can be difficult, particularly when the cuttings are very fine. The problem with a sandstone, for example, is to determine its *in situ* state of consolidation, because the drilling may dissociate it into its constituent grains. In this case, one has to rely heavily on the geophysical log, the driller's log and cores. Cuttings from coarse formations such as gravel or conglomerate will be smaller than the formation grains, and here, the true identity of the formation may be obtained from other logs and from the observation that the bulk of the cuttings are crushed, angular fragments. Similarly, cuttings from crystalline rocks will *all* be angular and fractured, and the rock has to be identified from the larger cuttings or from a mineral percentile log.

7.3 Geophysical logging

The measurement of features in a borehole, or in the geological formation adjacent to the borehole, can be carried out by geophysical logging. A logging unit comprises a monitoring console, a set of tools and a winch with the necessary conductor cable. Each tool is designed to measure one or more variables as it is lowered down the borehole on the end of the cable. Measurements are sent as electronic signals from the tool, up the cable, to the console as a continuous record of the parameter being studied. In the past, the electronic signals were converted by the surface console directly to analogue traces or logs on a chart recorder (Fig. 7.5), but now many logging units store the signals in data loggers. The information in the data logger is digitized and can be recovered as a standard chart record, or can be used and manipulated in log analyses.

A logging unit capable of logging holes much over 500 m deep is heavy, principally because of the winch, and has to be truck-mounted. More portable units, however, are available (Fig. 7.6).

The range of geophysical tools now available is enormous, mainly because of their use in the oil industry, but a restricted range of tools is normally used in the water industry. The geophysical logs for hydrocarbon wells are used largely for formation evaluation (Pirson, 1963) and their quantitative analysis can be quite complex. The logs most commonly used in the water industry can be divided into three types: formation, structural and fluid logs. Their interpretation is usually qualitative in nature and rarely as sophisticated as for oil wells.

The formation logs include resistivity, self-potential (SP), gamma ray, neutron and gamma–gamma. The resistivity and SP logs are commonly run together as an electric log, which can be valuable in identifying lithologies, but suffers from a lack of precision and its limitation to fluid-filled, uncased holes. The deflections on an SP curve are generated at the junction between a permeable bed, an impermeable bed and the drilling fluid, so that the log can be used to detect boundaries between sands and clays. The resistivity logs measure the formation resistivity, which depends to some extent on the groundwater salinity, but is also characteristic of the rock type:

89

The field guide to water wells and boreholes

Fig. 7.5 Geophysical logs.

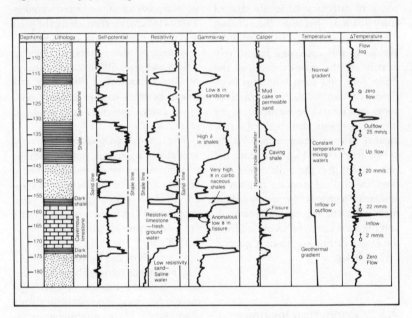

Fig 7.6 Portable geophysical logging unit.

Fig 7.7 Correlation using a gamma-ray log.

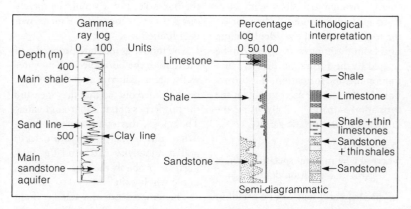

High resistivity   Crystalline rocks,
                    limestones etc.
                    Sandstones
                    Sands, unconsoli-
                    dated
                    Silts
Low resistivity   Clay and shale

The gamma-ray log measures the natural gamma-ray emissions from formations and can be invaluable in identifying various lithologies for comparison with a percentile log (Fig. 7.7). The natural gamma count generally reflects the shale content of sedimentary rocks because most gamma rays comes from radioactive potassium isotopes in clay minerals. The great advantage of this log is that it can be used in dry or cased boreholes.

The neutron and gamma–gamma logs use small radioactive sources to emit neutrons and gamma rays respectively. The use of such sources makes the logistics of logging much greater because their storage, transport and use on-site are subject to rigorous safe-

ty rules. These radioactive logs, therefore, are used much less frequently than the others.

The neutron tool emits high-energy neutrons, which lose their energy on collision with atomic nuclei in the formation, and eventually become captured. The lost energy is emitted as gamma rays. The greatest energy loss is when the neutrons hit hydrogen nuclei (protons), and therefore the energy loss and emission of gamma ray is dependent on the amount of hydrogen (as water) in the formation. The percentage of water in a saturated formation is the *formation porosity*, and so the gamma-ray emissions, measured by a detector on the tool, can be calibrated in terms of the porosity. Other neutron tools have detectors for slow or fast neutrons instead of gamma rays, but are based on the same effect of neutron capture by hydrogen nuclei.

The gamma–gamma tool is used to measure the formation density. Medium-energy gamma rays are emit-

91

ted from a source in the tool and lose energy through collisions with electrons in the formation. A detector on the tool measures the residual gamma rays from the emitter and these can be related to the formation density. In a monomineralic formation, this formation density can be used to obtain the formation porosity (Ø) by using the known mineral specific gravity.

$$Ø = \rho 1 - \rho 2 / \rho 1 - 1$$

where $\rho 1$ = mineral specific gravity;
$\rho 2$ = formation density.

The structural logs can be useful in formation evaluation, but are primarily run to check on the physical status of the borehole. The most common structural logs are the caliper, the casing collar locator and the closed-circuit television (CCTV) inspection. The basic caliper tool is lowered to the bottom of the well, where three (or four) spring-loaded arms are extended by servo-motor. The tool is then pulled to the surface, and the diameter of the hole is measured as twice the average radius measured by the arms. Sophisticated caliper tools can record the radii of the different arms to give a more complete geometry of the hole. The caliper tool identifies zones of caving, zones of thick mud cake, fractured casing and fissures in indurated rocks. The casing collar locator detects small electric fields generated at junctions at different metals, and can be used to detect casing joints or faults in casing. The CCTV gives a visual display of the well walls, but it does suffer from the fact that the picture quality can be destroyed by cloudy water. The visual inspection is good for detecting crack-

ed casing, fissures in rocks or debris in the borehole, but it should be merely one of a suite of logs. CCTV on its own is of limited use.

The fluid logs measure properties of the fluid in the borehole and, in water wells, are usually run after the well has been cleaned out, so that they measure the properties of natural groundwater. These logs, the temperature, conductivity, differential temperature (ΔT) and conductivity (ΔC) and the flow logs, are probably the most useful logs in the whole suite.

The temperature of a static column of water in a borehole will rise with depth according to the geothermal gradient. Groundwater flow across or up a borehole will disturb this temperature gradient. Differential temperature (ΔT) measures the change in temperature over a fixed depth interval, so that, with a uniform gradient, ΔT is constant, but a temperature change caused by an inflow of water creates a peak on the ΔT log (Fig. 7.5). The conductivity log measures the conductivity of the water in the well and gives some indication of its quality. A change in conductivity (a peak on the ΔC log) will mark an inflow or outflow of water to the well. In a well where heads vary through the thickness of the aquifer, there will be flow up or down the water column. An impeller flow-meter can measure high flows, but at low flow velocities of a few cm/s, a heat pulse flow-meter may be needed. The interpretation of this full suite of flow logs can provide invaluable data on the hydraulic regime in the borehole and surrounding formations.

Logging of boreholes and wells

Table 7.1   Application of borehole geophysics

Type of log	Geophysical log	Cased/screened borehole	Uncased borehole	Mud/water-filled borehole	Dry borehole
Formation	Resistivity	No	Yes	Yes	No
	Self-Potential	No	Yes	Yes	No
	Gamma	Yes	Yes	Yes	Yes
	Neutron	Yes	Yes	Yes	Yes
	Gamma–Gamma	Yes	Yes	Yes	Yes
Structural	Caliper	No use	Yes	Yes	Yes
	Casing collar locator	Yes	No	Yes	No
	CCTV	Yes	Yes	Only in clean water	Yes
Fluid log	Flow-meter	Yes, but of limited use	Yes	Not in mud	No
	Temperature/Δ Temperature		Yes	Yes	No
	Conductivity/Δ Conductivity		Yes	Yes	No

Geophysical logs can be used to provide information to help in pumping test analyses or regional hydrogeological studies. They are also useful in checking completion details of wells in drilling contracts. A suite of logs also represent a base set of data against which future well performance or status can be measured. For all these reasons it is recommended that every borehole drilled should have a basic set of logs run on its completion. This set should include:

1. Gamma-ray, caliper and flow-meter.
2. Temperature + ΔT.
3. Conductivity + ΔC.

The limitations of the use of various geophysical logging tools in boreholes with or without casing, or containing water columns or not, are summarized on Table 7.1.

Geophysical logging is a specialized operation and should be done by a geophysical contractor or trained operator. On-site interpretation can be undertaken by a hydrogeologist through a visual comparison of the suite of logs, but a comprehensive analysis is better done by a geophysicist in the office. This comprehensive analysis is better with digitally recorded logs.

8

Water well development

The methodology of development is discussed below with respect to water wells, but the same methods can be applied to any other kind of borehole or well.

The action of drilling a well will invariably lead to some damage to the aquifer immediately adjacent to the structure, and result in a reduction in the well's potential performance. The primary purpose of well development is to repair the damage done to the aquifer and restore the well's performance. The secondary aim is to develop the aquifer itself by increasing the transmissive properties of the aquifer adjacent to the well to values actually greater than they were before drilling.

8.1 Well and aquifer damage

The deleterious effects of drilling fall into two categories—damage to the well face and damage to the aquifer matrix. The relative importance of the two categories will depend on the drilling technique being used.

The reciprocating piston action of a shell or drill bit in percussion drilling can seal a well face very effectively by smearing clay over the borehole wall, particularly if there are clay layers in the geological succession. In a fissured aquifer, wall-smear or the squeezing of clay into the fissures can cut off the entire flow of groundwater into the well. The surging action of the tool string may also force dirty water into the aquifer matrix but, because the level of water in a well during percussion drilling is usually lower than in the ground, the formation invasion is not likely to be extensive.

The build-up of a mud cake on the well face is a necessary part of direct-circulation mud-flush rotary drilling (Section 4.2.1), in order to support the borehole and to prevent the loss of drilling fluid from the borehole. This means that the mud cake prevents the flow of water into the borehole, that is, it effectively seals off the aquifer. The mud cake is a layer of firm clay, commonly about 10 mm thick, strongly keyed to the porous aquifer. The removal of such a cohesive layer can be very difficult and require violent methods. The high hydraulic heads maintained in a rotary-drilled hole will force some of the fluid from the drilling mud through the mud cake, because the mud is retained by the porous well face and a filtrate of cleaner water passes into the aquifer. This water may be cleaner than the drilling fluid, but it still contains fine material and can block the pores of the aquifer matrix in

The field guide to water wells and boreholes

a zone of invasion around the borehole (Fig. 8.1).

Reverse-circulation drilling usually uses water instead of mud-based drilling fluids but, though this water may be clean at the start of drilling, it quickly becomes contaminated by the finer material from the formations being drilled; these fines will be actively flushed into the aquifer under the high positive head imposed by this drilling technique. If the aquifer is a fine- to medium-grained sand, then it also may build up a thin but effective mud cake, and, if the formation being drilled happens to be micaceous, the mica flakes can totally blind a well face with a layer much less than 1 mm thick. Figure 8.2 shows the effects of aquifer blockage resulting from recharging a medium-grained sandstone with potable water containing a minute fraction of alum floc derived from back-washing filters at the treatment works supplying the water.

A borehole in unconsolidated aquifers has to be supported, while drilling, by temporary casing, mud cake or by the hydraulic head of the fluid column. This means that the per-

Fig. 8.1 Mud filtrate invasion of an aquifer.

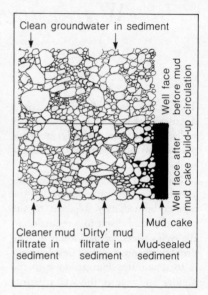

manent screen and pack have to be installed before any development can take place, and this results in the mud cake and damaged aquifer being separated from the well bore by a thick gravel pack or by a screen. The blinding of a well face by mud cake or

Fig. 8.2 Effect of blockage of a well face.

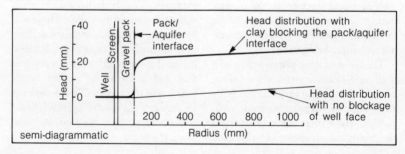

micaceous layer is serious, though curable by development, when the well face is open, but when the face is hidden by a gravel pack, the problem is much worse and development is not always possible.

8.2 Well development

The main problem in well development is the removal of wall-smear or mud cake and mud filtrate derived from bentonite-based muds. The methods of development are dealt with below, but clearly it would be best if the need for development could be avoided altogether. A relatively new mud technology avoids this bentonite problem altogether by replacing the bentonite with a biodegradable organic-polymer-based mud (Section 4.2). This polymer mud does reduce the problem of well development, but it does not remove it altogether. Mud cake and formation invasion by fines from the formation being drilled still takes place. Also, since the polymer mud is organic, it has to be removed totally from the well or it may act as food for bacteria and encourage infection of the well. The driller must ensure that the polymer is not left behind the blind casing, where development cannot be effective. Care also should be taken in the disposal of the water containing the polymer mud during clearance pumping, because its oxygen demand can be very high. The mud could de-oxygenate the water and kill fish if discharged into a surface water course. The addition of an oxidizer such as hypochlorite will 'break' the polymer mud and reduce the oxygen demand.

In a borehole with an artificial gravel pack, the development of the pack is extremely important in order to ensure that it is stable and fills the annulus, and to remove the fine fraction. Figure 8.3 is a photograph of a robust stainless-steel bridge slot screen, from a well in which bridging of the gravel occurred and cavities developed in the pack. Turbulent eddies in these cavities sand-blasted the back of the screen and perforated it in about three years.

8.3 Aquifer development

Aquifer development is merely the extension of well development to the stage where the fine fraction of the aquifer itself is being removed. There is no clear demarcation between the two stages of development.

Aquifer development will lead to a zone of enhanced porosity and hydraulic conductivity around the borehole, which will improve the

Fig. 8.3 Hole sand-blasted through a screen.

borehole performance. This increase in well efficiency is the result of a decrease in the velocity of groundwater flow in the zone adjacent to the well screen. The natural flow velocity of groundwater is low, usually less than a metre a day, and the flow is laminar. As groundwater approaches a pumping well, it has to flow through progressively smaller cross-sections of aquifer, and therefore, for a given well discharge, will flow at increasing velocities. The groundwater velocity one metre from a well will be four orders of magnitude greater than at a radius of 100 m. The groundwater velocity can be high enough for the flow to become turbulent, which leads to energy losses and has the effect of increasing the drawdown in the well for a specific discharge rate. This increase in drawdown is called the *well loss* (Section 9.3). The relationship between the groundwater velocity, the aquifer porosity and the hydraulic conductivity, is shown by a modified Darcy's equation:

$$v = Ki/\varnothing$$

where v = groundwater velocity (m/day); i = hydraulic gradient; K = hydraulic conductivity (m/d); $\varnothing$ = aquifer porosity (fraction).

An increase in porosity owing to development, will lead to a proportional decrease in v in this critical zone where turbulent flow is most likely to occur, and so will reduce the potential well-loss factor in the drawdown.

Aquifer development can improve the well performance but, because its effect is local to the well, it cannot produce additional groundwater resources from the aquifer as a whole, although it does allow the groundwater to be withdrawn more efficiently.

Successful development of a granular, unconsolidated aquifer will lead to the formation of a natural gravel pack (Section 3.2.1). Development could remove up to 40% of the aquifer material close to the screen but, because the effects of development will decrease away from the well face, the gravel pack will grade from a clean gravel against the screen, to the natural aquifer material (Fig. 8.4). Development will affect only the aquifer close to the well, usually within a metre of the screen.

A well in a fine, well-sorted aquifer should be equipped with a screen and artificial gravel pack (Chapter 3), and the development of such a well should be undertaken with care, because too vigorous a development could mobilize the aquifer material to break through the pack and lead to sand pumping. The opportunity for development in such an aquifer, apart from repairing aquifer damage, is in any case limited, because there is not a distinct finer aquifer fraction.

The development of a fissured aquifer may involve cleaning out fine debris from the fissures, for example, in the UK, removing Keuper Marl from fissures (neptunian dykes) in the Permo-Triassic sandstone aquifer. The most common fissured aquifer development, however, is the widening and cleaning of fissure systems in calcareous aquifers by acidization. This development is concentrated on

Fig. 8.4 Natural gravel pack development.

Fig. 8.5 Wall scratcher and surge block.

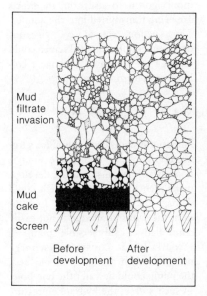

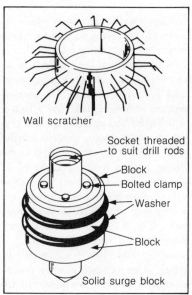

very restricted sections of the aquifer—the fissures, and its effect can extend a considerable distance, commonly tens of metres from the well.

8.4 Methods of development

Development relies on either physical or chemical methods to wash the well face and mobilize the material to be removed from the well. The physical methods include scratching, surging and jetting, while the chemical methods include polyphosphate dispersion and acidization.

8.4.1 Wall scratchers

Wall scratchers (Section 4.2) are circles of radiating steel, spokes which are strapped or welded at intervals along the screen/casing string as it is installed in a well (Fig. 8.5). The rings of the spokes act like a flue brush and scrape the well walls as the screen string is lowered down the well. They are not recommended for use in boreholes drilled in uncemented aquifers, because of the danger that they might cause caving.

99

8.4.2 Surge blocks and bailers

Surging a well is a process which attempts to set up the standard washing action of forcing water backwards and forwards through the material to be cleaned—in this case the screen, gravel pack and aquifer matrix. The simplest method of surging is to use a bailer, which acts like a piston in the well casing and will pull loose material into the well, where it can be removed in the bailer. Such a bailer, because of its loose fit in the casing and its flap-valve cannot push water back through the screen with any force. A solid surge block (Fig. 8.5) will provide a much more forceful surge. The block has several flexible washers, sized to fit tightly in the well casing, and is fastened either to a rigid drill stem on a rotary rig or in a heavy tool assembly on a percussion rig. The block is put below the water level inside the well casing, but not in the screen, and then rhythmically moved up and down on a stroke length of about a metre. Water is forced out through the screen on the down-stroke and pulled back through the screen on the up-stroke. The force of the surging is dependent on the speed of the strokes, but care should be taken to avoid damage to the screen by excessive force. Intermediate in force between a solid block and a bailer, is a surge block with a valve system which allows water to pass through on the down-stroke. Surging will bring material into the well, which will have to be periodically cleaned out by bailing. The development should start gently to ensure that water can move through the screen, before becoming more vigorous to ensure that the effects are being transmitted into the aquifer. The suction caused by vigorous development of a blocked screen could damage the screen by making it collapse or deform.

8.4.3 Surge pumping

Pumping systems can be used for surging a borehole. A submersible pump is not advised, because, during development, debris is pumped which could damage the pump impellers. In addition, the pump discharge is not easily controlled—a submersible pump is essentially a constant-discharge pump—and, if the screen is blocked, the pump could dewater the borehole. In such a case, the hydraulic pressure behind the screen can be great enough to collapse the screen. Screens collapsed in this manner are not common, but several cases have been reported and underline the rule that development should start gently to ensure that water can pass through the screen.

Overpumping, or pumping at discharges as great as the borehole will yield, is a common method used for development, but this may be ineffective. Pumping at a constant rate, however high, will remove loose material but, because there is no surging, could stabilize a situation of partial development. Development by pumping must be in a manner designed to induce surging; this can be done by using a turbine pump without a non-return valve. The pump is switched on

Fig. 8.6 Air-lift pumping arrangement.

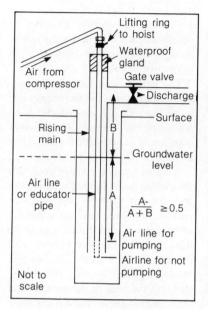

for a few minutes discharge, then switched off again and the sequence repeated. Water is pulled into the well by the pumping, but when pumping stops and when the rising main empties back into the well, water surges back through the system.

Air-lift pumping is very suitable for well development because there are no parts to be damaged by the debris drawn into the well. In normal air-lift pumping (Fig. 8.6), the air line from a compressor is inside the rising main, with the end of the air line at such a depth that at least 50% of the length of the air line is submerged. The rate of pumping can be controlled by the volume of air passed down the air line. When pumping has started satisfac-torily, it can be stopped and started at short intervals by shutting off air from the compressor; this will induce surging. A gate valve on the discharge pipeline can also be used for surging, by allowing pressure in the pipe system to build up, until either the air forces water out of the rising main and starts bubbling up the well, or it is held in the rising main by the hydrostatic pressure in the well. In either case, when the gate valve is suddenly opened, there is a violent release of pressure and water is pulled through the screen as pumping begins. When the valve is closed, water is forced back through the screen as pressure builds up again.

A further alternative is to set the air line just a couple of metres from the bottom of the rising main, then the pumping/non-pumping cycle can be set up by raising the air line inside the rising main (pumping), and lowering it below the rising main (non-pumping). The rapid alternation of pumping and non-pumping phases is a vigorous method of well development and is commonly called 'rawhiding' in the USA.

8.4.4 Jetting

Jetting is the washing of the well face with high-pressure jets of water. On a rotary rig, water is pumped, using the mud-pump, down the drill string to a jetting tool fixed to the end of the string (Fig. 8.7). The jetting tool has three or four nozzles at right angles to the drill string, so that the water jets are directed at the screen slots or borehole face. The pressure of the jets will de-

101

Fig. 8.7 Diagram of jetting head.

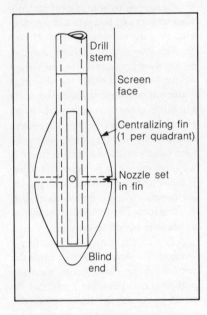

Drill stem

Screen face

Centralizing fin (1 per quadrant)

Nozzle set in fin

Blind end

pend on the depth of the well, but usually will be between 0.5 and 1.5 MPa. The nozzle diameter is usually about 4 mm, and water will be pumped at a rate to produce a jet velocity of at least 30 m/s.

The jetting head should have centralizers attached, and the nozzles can be on the ends of stems which can be adjusted for screens of different diameter. The nozzle orifice should be as close to the well face as possible.

A jetting head on the drill string of a rotary rig, can be rotated as it is raised and lowered past the section to be cleaned. The jetting should be done in short sections, and should start at the top of the screen or open borehole and progress to the bottom. This method of development is very effective in cleaning the well or screen face and removing mud cake, but it is less effective in restoring damaged aquifer matrix. The energy of the jets is quickly dissipated in the aquifer matrix and,

Table 8.1 Jet performance data

Jet no.	Nozzle diameter (mm)	Jet velocity (m/sec)	Discharge rate	
			(litre/s)	(Imperial gallon/min)
3	3	30	0.64	8.4
3	3	60	1.28	16.9
3	4	30	1.13	14.9
3	4	60	2.26	29.8
3	5	30	1.78	23.5
3	5	60	3.56	47.0
4	3	30	0.85	11.2
4	3	60	1.70	22.4
4	4	30	1.51	19.9
4	4	60	3.02	39.9
4	5	30	2.36	31.2
4	5	60	4.72	62.3

unless some pumping system is incorporated with the jetting, the flow of water is into the aquifer, so driving fines further into the matrix. The pumping rate must be greater than the jetting rate, so that there is a flow of water from the aquifer into the well.

Jetting a percussion-drilled hole has to be done by a jetting rig which is separate from the drilling rig. A jetting tool can be lowered down the borehole on a flexible pressure hose, down which water is fed from a surface pump. The principle is similar to that used on a rotary rig, but there are no facilities for rotating the head, so that washing may be incomplete.

8.4.5 Polyphosphate dispersants

The cohesiveness and plasticity of a clay can be broken down by dispersants, the most common of which are the polyphosphates such as Calgon. The chemical is often supplied as a granular hygroscopic material which has to be dissolved before it is added to the well. When mixing the solution, the solid is added to warm water a little at a time and stirred with mechanical mixers or it can set into an intractable glacial mass. Dosage is calculated at about 10–50 kg/m³ of water in the well. The lower concentrations may be used to help in cleaning fine materials for coarse packs, while the higher concentrations would be used to remove intractable mud cakes from borehole walls. The chemical acts by causing the individual flakes of the clay minerals to repel each other and so

break up clay granules. The clay in the mud cake, infiltrated mud, or mud in the original rock matrix, becomes less cohesive, more dispersed and is more easily removed by washing.

The polyphoshate should be left as long as possible, preferably at least overnight, to have time to permeate the formation and act. Jetting or surging can help the process and this combined use of physical and chemical methods is recommended.

8.4.6 Acidization

In the water industry, acidization involving the use of hydrochloric acid or sulphamic acid is used for the development of wells mainly in carbonate aquifers.

Hydrochloric acid HCl, also called muriatic acid, operates by dissolving the calcium carbonate from wall-smear or from drilling debris forced into fissures while carbonate aquifers (limestones or chalk) are being drilled. The acid is supplied as a solution of hydrochloric acid in water, and can be specified by its concentration, specific gravity or degrees Twaddell (Fig. 8.8).

The acid solution is usually supplied at a concentration of about 30 degrees Twaddell, containing a suitable inhibitor to prevent the acid unduly corroding the steel wall-casing and screens on the pump fittings (Williams *et al.* 1979, p. 92). The use of strong acid can be a hazardous procedure and should only be undertaken by professional competent organizations.

The acid solution is injected into the

The field guide to water wells and boreholes

Fig. 8.8 Hydrochloric acid concentrations.

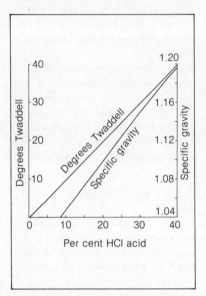

borehole, and then the well head is sealed with a cap equipped with a safety valve. The dissolution of limestone by the acid generates carbon dioxide, which builds up a pressure against the well head and forces the acid into the formation. Sufficient acid has to be added to the borehole to maintain the strength needed to dissolve the carbonate debris. The amount used will depend on the volume of water in the borehole and the length of the well face to be developed. The dilution available in the well and its immediate surrounds can be considerable, so that in a typical English Chalk well, between ten and twenty tonnes of acid may be used.

In a fissured karstic aquifer, the acid can develop the aquifer by cleaning and widening the fissures for a con-

siderable distance away from the well face. This ability of the acid to travel along fissures also means that care must be taken during acidization to avoid contamination of nearby pumping wells. The pumps in all neighbouring wells, to a radius of at least 100 metres, should be switched off during the acidization.

A further problem with acidization is the disposal of the spent acid; most pollution control agencies in England (attached to the water authority) will set limits on the pH and chloride content of the effluent to be discharged. These limits should be ascertained before acidization begins. The spent acid is withdrawn from the borehole, slowly at first, and put in a storage tank where its pH can be neutralized and its chloride level reduced. The pH and chloride levels should be monitored by on-site meters. If necessary, this 'strong' spent acid may have to be sent to a hazardous waste site. Later, as the spent acid becomes more dilute and within the discharge limits imposed, it can be put into a storm water or foul sewer. Acidization of carbonates produces carbon dioxide, and measures should be taken to avoid this heavy gas filling any enclosed spaces, where it could asphyxiate people.

Sulphamic acid is supplied as a granular material to be dissolved in warm water on-site, to give a strong acid solution. A saturated solution of sulphamic acid at 60 degrees Fahrenheit will contain about 20% dry acid by weight and have a pH of about $0 \cdot 4$. The reaction between the acid and carbonates is less vigorous than with hydrochloric acid, and consequently,

104

Table 8.2 Choosing a well development programme

1. Is the aquifer a limestone? YES—then acidization is likely to be the method of development:
   Check
   a) Have other wells in the area been acidified? If so, check the results to see if the method is effective.
   b) Does your contractor have acidization equipment and experience? If not then make sure he knows how to acidize safely.
   c) Is acid available locally? If not, go to option 2.
   d) Will the water authority permit disposal of the spent acid. Ensure the contractor will comply with their pollution control regulations.

2. NO—and the borehole will be drilled by mud-flush rotary methods.
   Then a mix of methods may be used—scratching, jetting, dispersants or surging.
   Ensure that the contractor has the necessary equipment, or chemicals and knows how to use them.
   Check that the contractor has either surge blocks or a surge pump and agree the method to be used.

3. NO—but the borehole will be drilled by percussion methods.
   Then development by surging and dispersants may be sufficient.
   Ensure that the contractor has the equipment and chemicals and knows how to use them.

4. Clearance pumping will be needed after all development work.
   Ensure the contractor has air-lift equipment or has, and is willing to use, a submersible pump.
   Check that he has equipment to remove wastes and spent chemicals and has a place to dispose of them legally. If permissible concentrations of spent chemicals are defined by the pollution control agency, then ensure that the contractor has the equipment to measure those concentrations.

residence times required for treatment are longer. Corrosion of metal fitments is less than with HCl.

The main advantage of sulphamic acid is its convenience and ease of use. the main disadvantage is its cost—it is much more expensive than HCl, and its use really can be justified only in isolated sites where transport is a problem, or at sensitive sites when tankers of HCl may present a hazard or be unacceptable. Table 8.2 summarizes the factors to consider when choosing a well development programme.

9

Well testing

Pumping tests of wells and boreholes are carried out for three main reasons:

1. To measure the well performance.
2. To estimate the well efficiency or the variation of well performance with the discharge rate.
3. To measure the aquifer characteristics of storativity, hydraulic conductivity and transmissivity.

These may be combined as an overall reason, namely to evaluate the well hydraulics of the pumping well and the aquifer characteristics of the exploited aquifer.

What is a pumping test? Under natural conditions, groundwater flows through aquifers, from areas of high hydraulic head to areas of low head where the water discharges to the surface (Section 1.1.2). This natural flow is disturbed when a water well is pumped, and a pumping test is designed to cause such a disturbance under controlled conditions, so that its effects can be analysed to obtain values of well performance, well efficiency or the aquifer characteristics.

Different types of test are used (Section 9.2) but the most common types are:

1. Step drawdown test to measure the well efficiency, to measure the well performance and to measure the variation of well performance with the discharge rate.
2. Constant discharge tests to measure the well performance and to measure the aquifer characteristics.

The disturbance to groundwater flow is observed by measuring the changes in the water levels in an array of observation boreholes specially drilled around the pumping well (Fig. 9.1). The site measurements involved are those of time through the tests, water levels in the boreholes and the discharge rate of the pumped well.

The pumping lowers the water level in the well, and the resultant head difference between the aquifer and the well induces flow into the well. The flow into a well from an ideal aquifer is radial, and the effect of pumping spreads outwards radially as a cone of depression in the piezometric surface (Fig. 9.2). The need to observe the growth of this cone, and its possible interference with surface features such as streams, dictates the distribution of boreholes at each site.

The analyses of constant discharge tests are dependant on the comparison of the observed behaviour of the groundwater levels with the theoretical behaviour in a perfect aquifer, as defined by the Theis equation or its variants

Fig. 9.1 Borehole array for a test well.

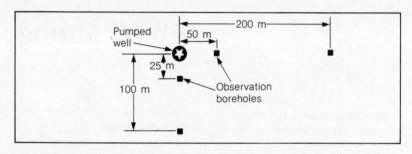

Fig. 9.2 Diagram of a cone of depression.

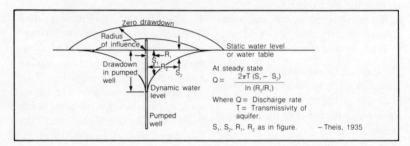

(Section 9.3). The traditional method of comparison is by graphical super-position of data curves over type curves, or by the analysis of the data curves (Kruseman & De Ridder, 1970). These graphical methods of analysis are being replaced or supplemented by computer analysis (Rushton & Redshaw, 1979). Pro-grammes are now available to model water-level responses to pumping, which take into account well storage, variations in aquifer storage, permeability and leakage as well as 'normal' aquifer hydraulics.

The analysis of the growth of the cone of depression is discussed in detail in many hydrogeological textbooks. The reader is referred to Davis & De Weist, 1966; Freeze & Cherry, 1979; Todd, 1980 or Hamill & Bell, 1986. A valuable handbook on pumping-test analyses is the Bulletin by Kruseman and De Ridder, 1970.

The study of well and aquifer hydraulics by pumping tests can be supplemented by geophysical logging (Section 7.3). Pumping tests are used to analyse the groundwater flow regime beneath a site, and to indicate the potential performance of the well. Geophysical logging can be used to observe directly the condition of the well, to measure various features of the flow and quality of the groundwater, and to measure some aquifer properties.

9.1 Design of pumping tests

The measurements that one can make during a pumping test are:

1. The discharge rate.
2. The time since pumping started or stopped.
3. The water level in the pumped well and any observation wells.
4. The temperature and quality of the discharge water (Chapter 6).

Other factors which may affect or be affected by the pumping are:

1. Rainfall.
2. Tidal fluctuations.
3. Barometric variations.
4. Water level in surface ponds.
5. Flows in surface streams.
6. Water levels in other boreholes nearby.
7. Engineering structures.

The variables which can be controlled during a test are:

1. The discharge rate.
2. The duration of the test.
3. The number of observation wells.
4. The number of observations.

The options open to vary the design of a test fundamentally are therefore quite limited. The two most commonly used tests are the step drawdown (variable discharge) test and the constant discharge test. Several other types

of tests which are used much less frequently are summarized in Section 9.1.3, but the reader is referred to more specialized texts for details.

A problem common to all tests is the disposal of the discharged water. In tests of shallow aquifers, this water may infiltrate back into the aquifer and interfere with the test results. It is recommended that the discharge water is conducted either by pipeline or lined channel to beyond the range of influence of the test or to a nearby water course. A minimum distance of 200m is suggested.

9.1.1 Step drawdown tests

The primary aim of a step drawdown test is to evaluate the well performance rather than the aquifer performance. The drawdown in the pumped well is measured while the discharge rate is increased in steps. The discharge rate in each step is kept constant through the step, and the change in rate between steps is make as quickly as possible. Observation boreholes are not used, though measurements in such wells, during a test, may be useful in designing future constant discharge tests.

The factors to be decided in the design of a step test are:

1. The number of steps.
2. Duration of steps.
3. Discharge rate of steps.
4. Whether the steps should follow each other without a break.

A step drawdown test measures the variation of specific capacity with discharge rate, data which can be analysed to differentiate the proportion of the drawdown which is aquifer loss and that which is well loss. The test data can also give the value of the aquifer transmissivity and an indication of the storativity. The data on the specific capacity of the well are of great practical value, because they give a basis on which to choose the pump size and pump setting for the well in long-term production.

An analysis of a test (Section 9.3) involves the plotting of the specific drawdown (the reciprocal of specific capacity) and the discharge rate for each step, that is, each step produces one data point on the graph (Fig. 9.3). In order for the relationship between specific capacity and discharge rate deduced from such analysis to be valid, a minimum of four steps are required, because three points do not adequately define a trend. Five or six steps is the optimal number, because above that number the logistics of the test become difficult.

The most important feature of the step test is the number of steps rather than their lengths, but each step should be long enough to allow well-storage effects seen at the beginning of each step to dissipate. The steps are most commonly between 60 and 120 minutes long. There is little advantage in extending the steps beyond 120 minutes, but it does help the interpretation if the steps are of equal length. The length of time taken for each step is measured on a log scale, as if each step were a separate constant discharge test.

The maximum discharge rate of the test should be as great as the well or pump will give and, in either case—if possible—greater than the expected long-term production rate. The discharge rates for the other steps have to be chosen to give equal discharge increments through the test. These rates will have to be chosen before the test, and the pumping equipment should be calibrated during the equipment test,

Fig. 9.3 Step drawdown test data. To show that more than three steps are needed to draw a valid specific drawdown/discharge curve.

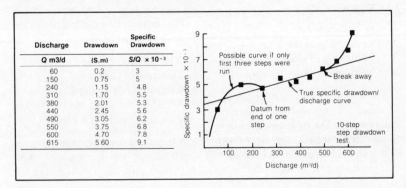

to give the required rates. At the beginning of each step, the discharge rate is set and then left; one should not try to adjust the rate to an exact reading. The discharge is monitored continuously through the test if possible.

The discharge rates through a test should start at the smallest and increase with successive steps. This rule is merely for ease of running the test and the test analysis, because modern analytical methods can take into account either decreases or increases in discharge rate.

The steps in a normal test follow each other without a break, but there is an argument that, to obtain accurately comparable test data, each step should be run as an individual constant discharge test. In practice, this would add inordinately to the length and cost of the test, with little demonstrable increase in data quality. The analyses of step tests do take into account the effects of preceding steps through a test using the theory of superposition.

9.1.2 Constant-discharge tests

The aim of a constant discharge test is to obtain data which can be interpreted by comparison with the behaviour of type aquifers predicted by Theis, Boulton or Thiem. The purpose of the test is to assess the well performance, to obtain values of the aquifer transmissivity and storativity, and to obtain some indication of the aquifer geometry in order to predict long-term aquifer and well performance.

The test programme has three distinct parts:

1. Pre-test observations.
2. Pumping test.
3. Recovery test.

The pre-test observations cover all those factors to be measured during the test, and should be made for a sufficient length of time in order to establish a trend or base level at each observation point. For example, if the water level in an observation well is affected by tidal influences, then a continuous record of the water level will be required for a period as long as the test, so that measurements taken during the test can be corrected. In a borehole where the water level is stable, measurements every few hours over a day, may be sufficient to establish the static water level. In a situation where the well to be tested is used for a water supply, and cannot be switched off for more than a few hours—for example in a factory—then after it is switched off prior to the test, the water-level recovery must be measured continuously. This is so that if the recovery is not complete, the data from the test can be corrected for incomplete recovery (Fig. 9.4).

An important part of pre-test observations is a short equipment test after the pump and all measuring equipment have been installed. The test pump is switched on to ensure the pump and discharge measuring devices are working. The water level in the well is measured during the few minutes pumping, to ensure that the discharge rate for the test does not exceed the capacity of the well. This equipment test

The field guide to water wells and boreholes

Fig. 9.4 Diagram to show correction of water levels for incomplete recovery.

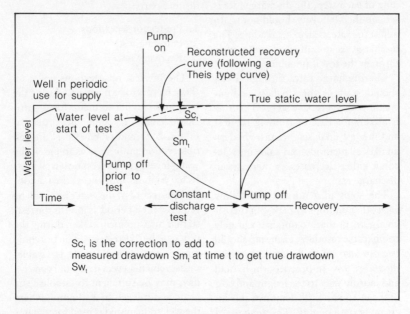

Sc$_t$ is the correction to add to
measured drawdown Sm$_t$ at time t to get true drawdown
Sw$_t$

should be run early enough to leave time for recovery before the main test.

The discharge rate and duration of the pumping test itself are decided before the test begins. The discharge rate normally is at least as high as the long-term production rate required from the well. If the equipment test or an initial step drawdown test has shown that such a rate cannot be sustained, then lower rate which can be maintained for the test period is chosen. The water level in a pumping well is very susceptible to changes in the pumping rate; changes in groundwater levels caused by discharge variations have been attributed, mistakenly, to aquifer variations in many tests. A continuous record of the discharge rate through a test is desirable to avoid such

misinterpretations; a daily measurement is not sufficient.

The length of a constant discharge test will be a balance between available budget and the need for useful data from the test. The effect of pumping spreads with time, but the rate of spread decreases logarithmically with time. At a fixed point such as an observation well, the rate of drawdown decreases with the log of time, according to the Theis equation. This means that there is a decrease in the 'rate of return', in terms of useful data, in extending a test period. This can be shown by a Theis curve, on which 100 minutes pumping will give two log cycles of data curve, and yet to double the length of that data curve would require a further 9900 minutes, almost 1

Fig. 9.5 A Theis curve to show a decrease in useful data as a test is extended.

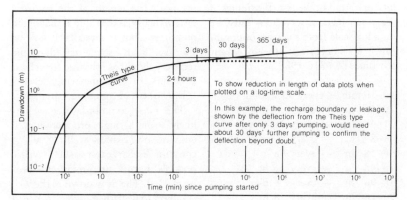

To show reduction in length of data plots when plotted on a log-time scale.

In this example, the recharge boundary or leakage, shown by the deflection from the Theis type curve after only 3 days' pumping, would need about 30 days' further pumping to confirm the deflection beyond doubt.

week of pumping (Fig. 9.5).

The length of test depends on a judgement of when the rate of return is not worth the expense, and this judgement will be site-specific to some extent.

Many short, constant discharge tests called yield tests are run for periods of a few hours, particularly on completion of single isolated boreholes for farms or small factories. The tests are run to show the client that the borehole is 'successful' and usually the measurements are minimal: commonly just the discharge rate and final drawdown. Such a test is better than no test, but it is of very limited value—it provides a value of specific capacity at one discharge rate. Even in a yield test, the water levels in the well should be measured accurately through the test, so that the results can be interpreted. In any case, a better test (if the test period is restricted to less than 12 hours) would be a step drawdown test, for this would give the client much more information about his well, for the same expenditure.

The recommended minimum length of a constant discharge test for a well which has no observation boreholes, or where the observation boreholes are within 100 m, is 24 hours. Where observation wells are more than 100 m from the pumped well, or in a water table aquifer where response time is slow, a three-day test is advised. A five-day test may be advised in sensitive situations where derogation effects on neighbouring wells need to be measured.

A test period beyond five days can only be justified where special conditions are present. These could be where the features to be studied take a long time to become evident or require time to be measurable; for example, leakage from a surface stream or inter-aquifer reactions. Constant-discharge tests involving more than one pumping well—group tests (Section 9.1.3) are used to test wide areas of aquifer, and require extended periods of pumping.

The variation of drawdown in an aquifer with the log of the duration of the pumping, means that, to get an

113

even spread of measurements for the interpretation of the test, the timing of the water-level observations must be on a log scale from the start of pumping. A suggested measurement schedule (BS 6316 1983) is:

0–10 min	— every 30 seconds
10–60 min	— every 5 min
1–4 hr	— every 15 min
4–8 hr	— every 30 min
8–18 hr	— every hr
18–48 hr	— every 2 hr
48–96 hr	— every 4 hr
96–168 hr	— every 8 hr
168 hr	— every 12 hr

This is a rigorous schedule in practice and should be considered as a standard to aim for.

On cessation of pumping, the groundwater levels will recover back to the static water level, following a time/drawdown curve which is the converse of the drawdown curve. The water levels should be measured from the cessation of pumping, on the same

kind of log time-scale as in the pumping test. The length of the recovery test, in theory, is the same as that of the pump test, but in practice it will be shorter. It is suggested that the recovery should be followed for at least 24 hours, or until the water level is within 10 cm of the static water level.

The most common water well is likely to be a single production well with no observation wells at all. A constant-discharge test of such a well is possible, but its interpretation is problematical. The drawdown in the well is subject to well losses, which can lead to an underestimation of aquifer transmissivity, unless the observed drawdown is corrected for the well loss (Fig. 9.6). The well loss can be determined by a step drawdown test, and therefore it is recommended that the testing of isolated wells includes both step drawdown and a constant-discharge test.

A new water well within a well field can be tested by using existing wells as observation boreholes. The planning

Fig. 9.6 Illustration of the correction of a Theis curve for well losses.

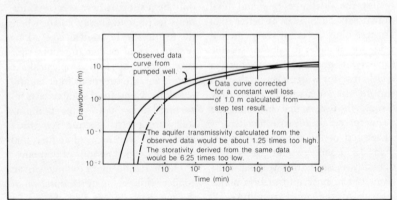

114

of a test in such a well field may be complex, because of the need to ensure that all surrounding wells are switched off or operating at a steady state.

A purpose-drilled test well will have satellite observation boreholes drilled with it (Fig. 9.1). Each observation borehole will act as an extra observation point in the aquifer, and so improve (in theory) the analysis of the test. Observation boreholes can be almost as expensive as the test well to drill, so budget constraints usually limit the number of observation boreholes. Commonly, a cruciform array of four observation boreholes is drilled in two pairs, at radial distances from the pumped well of about 25 and 100, and 50 and 200 m, along two radii at right angles to each other. This array enables the aquifer characteristics to be measured and an indication of the aquifer geometry to be obtained, and may be considered optimal. The numbers of observation boreholes may be reduced but, if at all possible, at least one observation borehole should be drilled with each test well, in order to obtain reliable pumping-test data.

9.1.3 Other tests

There are several types of test which are used less commonly than the constant-discharge and step drawdown tests. These tests include:
Group pumping test. A test designed to evaluate the aquifer behaviour over a wide area and to provide calibration data for mathematical groundwater models. Such a test involves simult-
aneous constant-discharge tests run on several water wells for a sufficient period to allow their cones of depression to coalesce.
Recharge test. A test where water is pumped into a borehole instead of out of it. The analysis of such a test is the same as for a constant-discharge test, except the discharge rate is negative.
Constant head test. A recharge test commonly used in site investigations for engineering works, in which the rate of recharge is adjusted so that the hydrostatic head in the borehole is kept constant.
Variable discharge test. A test in which the drawdown is kept constant and the discharge varied. This test is difficult to carry out using a variable discharge pump, but can be used to test a naturally overflowing (artesian) well.
Slug test. The commonest method of carrying out a slug test is to lower a large cylinder (the slug) on a line into the water column in the well, so that the water level is instantaneously raised a few metres. Alternatively, the slug is withdrawn to produce a drawdown and the recovery is monitored.
Packer test. The permeability of an aquifer very commonly varies through the aquifer thickness. Packer tests are one method to measure this variation. A limited thickness of aquifer is isolated between inflated packers (Section 6.2), and a constant discharge or recharge test carried out for that section. The packer assembly is then moved to a different section of the aquifer and the test is repeated. A profile of the permeability variation in the aquifer can be built up by repetitive testing.

9.1.4 Recommended pumping-test programmes

Each pumping-test programme will be site-specific, but the following guidelines, summarized in Table 9.1, cover most situations.

When less than twelve hours are available for the tests, then the best test is a step drawdown test which gives information on well performance, specific capacity and an estimate of the aquifer transmissivity and storativity. A short constant-discharge test to give the specific capacity of a borehole is not recommended, as it gives much less information than a step drawdown test (p.87). The step drawdown test could be supplemented by a constant-discharge test if a further 24 hours were available, to give more certain values of the transmissivity and storativity as well as indications of the aquifer behaviour.

The best general-purpose test programme is a step drawdown test followed, after recovery, by a 72-hour constant-discharge test and a final recovery period. These tests give full information on the borehole and the aquifer, provided that the aquifer is reasonably homogeneous.

Examples of sheets suitable for recording data from the various types of pumping tests are given in Tables 9.2 and 9.3.

9.2 Pumping-test equipment

Electric submersible pumps are now the most widely used pumps for testing high-capacity wells. These pumps are virtually constant-discharge pumps, and variations in discharge have to be achieved either by a gate valve on the discharge line, or by diverting part of the flow back down the well through a bypass system. The electric submersible has almost replaced the vertical shaft turbine pump powered by surface-mounted diesel or electric motors.

Clearance and development pumping of a new well is best done by air-lift pumping, because the debris in the water could damage a turbine pump. The same air-lift system might be used for subsequent pumping tests but it is not recommended, because the discharge rate is much more difficult to control than with a turbine pump.

With low-capacity wells, reciprocating pumps may be used for pumping tests. In developing countries these pumps have the advantage of being simple to maintain, requiring very few spare parts and needing neither diesel generator nor compressor as they can be run off a percussion rig.

The test pump has to be chosen, using manufacturer's pump performance curves, to be able to cope with the discharge rate and total lift anticipated during the test (Fig. 9.7). The power source then has to be matched to the pump, for example, with an electric submersible pump, the electricity supply has to be able to provide sufficient current to start the pump. The current needed to run the pump may be only 20% of that needed to start it, although various types of transformer can be used to reduce the starting current to two or three times the running current.

Table 9.1 The choice of a pumping-test programme

1. Are you testing a water well? — YES

    Then the *minimum* test is a six-hour constant-discharge test to give a specific capacity. If possible go further.

    Are there observation boreholes?

    YES   NO

Then the minimum recommended test is a step drawdown test to give specific capacity, transmissivity, well losses and well efficiency.	The step drawdown test is essential

NO

A constant-discharge test at least 72 hours long is recommended.

Follow step test with a constant-discharge test if more than 24 hours are permitted. A 24-hour test is possible but a 72-hour test is better. Data will need correcting for well losses from the step test.

Are there other wells or surface waters that the pumping may affect? If so, extend the constant-discharge test to at least 1-week duration.

Is the well artesian? If so, consider using a falling head test.

Then: 2. Do you need pumping-test data to validate a mathematical model or test a major resource scheme? Then a long-term group pumping test should be planned.
3. Are you undertaking site investigations for civil engineering purposes? Then you may use constant head, slug or packer tests.
4. Do you need measurements of *in situ* permeability or water quality from restricted aquifer sections? Then a packer test is needed.

117

The field guide to water wells and boreholes

Table 9.2 Pump-test record sheet—discharge period

Project	Date		Sheet		
Test site name			Grid reference		
Observation point name			Grid reference		
Distance (r) from pumped well (m)					
Site sketch			Type of test		
			Comments		
Time	Time (t) since pumping began (min)	t/r^2	Discharge rate (m^3/d)	Water level below datum (m)	Drawdown (m)

Table 9.3 Pump-test record sheet—recovery period

Project	Date	Sheet			
Test site name			Grid reference		
Observation point name			Grid reference		
Distance from pumped well (m)					
Discharge rate during pumping (m³/d)					
Static water level (m) below datum					
Site sketch			Type of test		
			Comments		
Time	Time (t) since pumping began (min)	Time (t') since pumping stopped (min)	t/t'	Water level below datum (m)	Residual drawdown (m)

Fig. 9.7 Examples of pump performance curves. For a demand of 15m²/hr against a head of 150 m, the series A, 15 HP, 23-stage pump would be chosen from this selection. The series B, 20 HP, 12-stage pump may do the job, but is at the top of its performance curve and would fail to deliver the discharge if the head changed much. After Godwin Pumps Ltd.

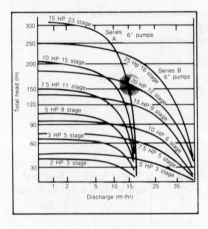

The measurements taken during the tests are: time, water levels and discharge rates. Modern digital wrist-watches with stop-watch functions are now adequate for any test.

The water levels in the pumped well and observation boreholes are most commonly measured using tapes with electric contact gauges (water-level dippers). The tapes are useful in that the electrical contact gauges can be repaired or replaced in the field if damaged or lost (Fig. 9.8). It is recommended that the water levels to be observed are monitored continuously using water level recorders. Float-activated clockwork chart recorders (Fig. 9.9) are the traditional type, but pressure transducers connected to electronic data loggers are beginning to replace them. The advantages of the pressure transducer systems, now that their price is more competative and their precision acceptable, is that they are less prone to failure than chart recorders, one data logger can be used for several observation boreholes simultaneously, and the water level/time data can be recorded in digital form, ready for immediate interpretation. The disadvantage of transducers is that they have to be set at a fixed depth below the static water level, and have a predetermined pressure range related to the expected

Fig. 9.8 Electrical contact gauge: field repair to damaged tip.

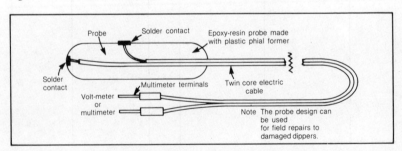

Fig. 9.9 Clockwork chart recorder for water-level measurement.

drawdown. This range may not be able to be predetermined if the expected drawdown cannot be estimated. A useful method of water-level measurement, particularly for the pumping of wells with deep water levels, is the air-line. The open end of the air-line tube is set just above the pump or below the deepest expected pumping water level. The water level is measured by pumping air into the air line and measuring the maximum air pressure obtainable. The water level in metres above the end of the tube is the air pressure in atmospheres × 0·0968.

The measurement of discharge rates in pumping tests in the UK is usually by a weir tank with a thin plate weir (BS 3680 Part 4A) (Fig. 9.10). The weir is usually a V-notched type, but rectangular weirs are sometimes used. The discharge rate is proportional to the head over the weir and a function of the shape of the weir. The accuracy of the flow measurements will depend upon the weir tank being set up in accordance with the BS specification. It is

Fig. 9.10 Diagram of a weir tank.

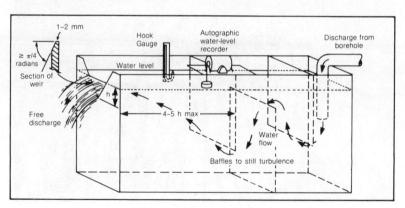

recommended that the water level in the tank should be monitored continuously by a recorder on the tank, and that periodic measurements should be taken by an accurate hook gauge. Discharge values for a range of head measurements over 90° weirs are given in Appendix A.IV.1.

Flow measurement by orifice plate is used extensively outside the UK. This is more convenient than the weir tank because its component parts are less bulky to transport (Fig. 9.11). The system comprises a horizontal pipe, with a plate set in its end in which there is a circular orifice of smaller diamter than the pipe. The restriction creates a back pressure which is measured by a manometer tapped into the midline of the pipe. The manometer reading is

dependent on the flow rate and the ratio of the pipe ID and the orifice diameter (See Appendix A.IV.2).

Flow-meters may be set in the discharge line to measure the flow rate, but generally they are accurate only over a defined range of flows, and care must be taken to ensure that the correct gauge is used for the test. An on-line flow-meter should not be used in a step drawdown test or any other variable discharge test.

9.3 Well and aquifer hydraulics from pumping tests

The methods of pumping-test analysis are described in many textbooks (p.86). Here we consider only the basic

Fig. 9.11 An orifice weir in operation.

principles behind the common methods of analysis, so that the reader can select the most appropriate one for any situation. The basic analyses of well and aquifer hydraulics by pumping tests all depend on an idealized aquifer model and make several assumptions incuding:

1. The aquifer is infinite in extent.
2. The aquifer is homogeneous, isotropic and of uniform thickness.
3. The piezometric surface or water table is horizontal.
4. Groundwater flow is horizontal and uniform through the whole aquifer thickness.
5. The well discharge rate is constant.

It is important, when analysing pumping tests, not to forget these assumptions and try to apply analytical methods to inappropriate situations.

The first breakthrough in the analysis of the cone of depression was by Thiem (1906), who assumed that after a long period of pumping the cone of depression is in a steady state. Thiem showed that, at steady state, the drawdown in the cone is proportional to the log of the radius from the pumped well. The Thiem equation which expressed this relationship with respect to discharge and aquifer transmissivity, enabled the aquifer transmissivity to be determined for the first time by a constant-discharge test. Storativity does not enter into the equation, because at a steady state there is no change in storage conditions in the system.

The Thiem equation can not take into account the behaviour of the groundwater system during the *growth* of the cone of depression. A second breakthrough enabled this growth or transient element to be considered. Theis (1935) developed his equation for non-steady state flow to a well—which incorporates transmissivity, storativity and the duration of pumping. In addition to the standard assumptions, Theis also assumed that the aquifer was confined, the well was infinitely thin and that water was released instantly from storage as the head declined.

The Theis equation cannot be solved explicitly, but graphical methods have been developed for the analysis of pumping-test data, to give approximate solutions for transmissivity and storativity. The Theis analysis shows that, in an ideal aquifer, the drawdown at any point in a cone of depression will follow the form of a Theis type curve (Fig. 9.12)

The Thiem and Theis equations have been developed to explain the effect of a pumping well on an ideal aquifer. Developments in well hydraulic analysis since Theis have been concerned mainly with taking into to account the more common departures from the ideal state. The Theis analysis, for example, assumes that the aquifer is fully confined and that there is no leakage from, or into, the aquifer. Aquifers are rarely totally confined, and in nature, during pumping, some leakage will take place into the area of reduced hydrostatic head—the cone of depression—either from across or within the confining beds. The effect of leakage is to reduce the drawdown within the cone of depression. Leakage is detected in a pumping test by the observation that the drawdown data from observation wells follow a curve below the Theis type curve (Fig 9.12)

The drawdown/time curve during pumping from an unconfined (water table) aquifer will follow a Theis curve, provided that the drawdown in the pumping well is not a significant proportion of the saturated aquifer thickness. If the drawdown *is* significant, then the well behaves as if the aquifer transmissivity decreases through the test, and the data curve lies above the type curve. The drawdown/time curves of tests in water table aquifers, however, are commonly of sinusoidal form (Fig. 9.13) and this kind of curve could not be interpreted until Boulton (1963) showed that it was the result of *delayed yield*. Immediately after a pump is switched on and the water level in the well drops, water is released from aquifer storage by pressure release. The drawdown follows a Theis curve *but* the storage is the confined storage coef-

ficient (Section 1.1.2). Later, water is gradually released from storage by gravity drainage and the drawdown begins to follow a new Theis curve, with the storage being the unconfined specific yield. The time taken for the well behaviour to change from one state of storage to the other can vary from a few minutes to a few hours. The well response in the early phase of delayed yield is very similar to that in a leaky aquifer.

The Boulton analysis of delayed yield applies only to water table aquifers, but similar effects are observed in confined dual-porosity aquifers (Section 1.1.2). In fissured aquifers, for example, the fissures usually are highly transmissive but have low storage, while the rock matrix has a high storage but a low transmissivity. When pumping starts in such an aquifer, the fissures respond immediately, but then, gradually, as the water leaks from the matrix to the fissures, the matrix effects may come

Fig. 9.12 Pumping-test curves to how the effect of a leaky aquifer.

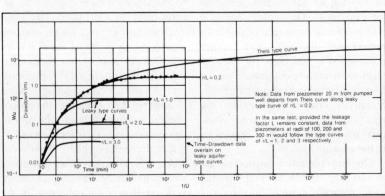

Fig. 9.13 Pumping-test curves to show the effect of delayed yield.

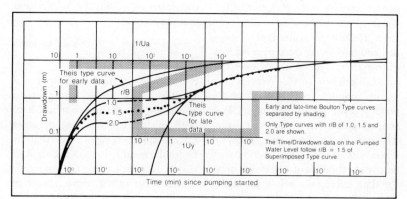

Fig. 9.14 Pumping-test curves to show the effect of recharge and barrier boundaries.

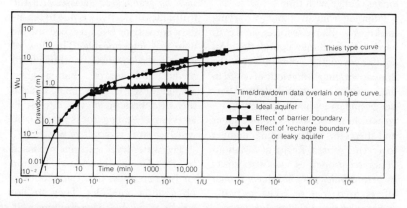

to dominate the well performance.

The Thiem, Theis and Boulton analyses all presuppose a uniform aquifer geometry, but this is rare in nature. An aquifer can be abruptly terminated by a geological fault, which acts as a barrier boundary and stops groundwater flow to the well from that direction. On a drawdown/time curve of a well in such an aquifer, the drawdown follows a Theis curve until

the fault is reached, and then the rate of drawdown increases and the data curve rises above the Theis curve (Fig. 9.14). The maximum effect of a straight fault is to apparently halve the aquifer transmissivity.

The opposite to a fault or barrier boundary is a recharge boundary. In a water table aquifer, the groundwater is commonly discharging to a stream and providing the base flow of that stream.

125

If pumping from a well lowers the water table below the elevation of the stream, then the base-flow increment from groundwater is intercepted, stopped and may be replaced by a net loss of river flow. This base-flow interception can result in surface streams drying up completely. The effect on the water level in the pumped borehole is similar to that from aquifer leakage. The drawdown/time curve follows a Theis curve until the boundary is reached by the cone of depression; then the data curve falls below the Theis type curve. When the recharge balances the discharge, a steady-state condition prevails and water levels no longer change with time.

Pumping from an aquifer of varying thickness will have features similar to barrier or recharge boundaries. An aquifer wedging out affects the drawdown/time behaviour of a well in the same manner as an ill-defined barrier boundary, while a thickening increases the transmissivity of the aquifer and affects the well in a similar manner to a recharge area. Regional variations in aquifer properties such as porosity or permeability may also have similar effects to barrier or recharge boundaries in terms of well performances.

The basic assumption that aquifers are homogeneous is also rarely valid. Unconsolidated sandy aquifers usually vary in grain size because they are commonly deposited in cyclothems, while groundwater flow in most consolidated aquifers is concentrated along joints or fissure planes. This heterogeneity does not affect the transmissivity of an aquifer as measured by a pumping test, but it does mean that the hydraulic conductivity varies through the aquifer and that the relation:

transmissivity = hydraulic conductivity × aquifer thickness

is *not* valid. In the Chalk aquifer of England, (a very fine-grained limestone) the fissure systems have been enhanced in the zone of water table fluctuation (Fig. 2.7), so that the hydraulic conductivity in the shallow Chalk may be orders of magnitude greater than in the same aquifer at depth.

The most extreme case of variable permeabilty is that of crystalline rocks such as granites or gneisses. All the hydraulic conductivity is due to secondary porosity or fractures, because the matrix is impermeable. The flow of water in such rocks is difficult to predict and analyse, but a great amount of work is now being done in this field because of the interest in assessing crystalline rocks as media in which to store radioactive wastes.

The water-level reactions predicted by the Thiem, Theis and Boulton analyses are those in the aquifer or in observation boreholes. They are not strictly applicable to the pumping well, because the well itself affects the reaction to pumping. The three analyses, for example, assume that the pumping well is infinitely thin, yet the water in the well, *the well storage,* affects the drawdown over the first few minutes of a pumping test. The pump is removing water from the well storage, so the drawdown is less than would be predicted from the aquifer characteristics alone. This can seriously affect a pumping-test analysis, because well storage affects the Theis curve in the first few minutes when it is at its steepest and most susceptible to change.

The velocity of groundwater flow in an aquifer under natural conditions is very slow but, near to a discharging well—particularly when the discharge rate is high, or if the flow is resticted to fissures or to thin, highly permeable zones—the velocity can be very high and the flow may become turbulent (Section 8.3). The situation may be made worse if the aquifer is damaged and has not been developed, or if the screen is inefficient. The direction of flow through a screen is at right angles to the axis of the well, and the abrupt change in flow direction needed for the water to move up to the pump will involve frictional losses of head against the casing string.

The head losses from turbulent flow or pipe friction increase the drawdown in the pumped well, and together are called well losses. Jacob (1946) suggested that the drawdown in a well comprises two parts—aquifer loss and well loss. The *aquifer loss* is that drawdown to be expected from the flow of water through the aquifer, as predicted by the Theis equation, and is proportional to the discharge rate. The well loss is the drawdown caused by all the turbulent and frictional effects in or adjacent to the well, and Jacob suggested that it is proportional to the square of the discharge rate. The validity of the well-loss proportionality has been disputed, but the squared relation does appear to hold in many cases, and the Jacob equation:

$$Sw = BQ + CQ^2$$

where Sw = drawdown in well; BQ = aquifer losses; and CQ^2 = well losses.

is the basis of most step-drawdown test analyses of well performance. These analyses began by using graphical techniques (Clark, 1977), but these are now supplemented by numerical methods of analysis (Rushton & Redshaw, 1979).

10
Well maintenance

Water wells deteriorate over the years and need periodic maintenance, like any other engineering structures. The main causes of the deterioration are incrustation of the screen and blockage of the gravel pack or aquifer. The deterioration is shown by reduced well performance, with a gradual reduction in the specific discharge (Fig. 10.1).

In order to quantify well deterioration, the well performance should be measured annually by a step drawdown test (Section 9.1.1). Regular monitoring in this way not only measures the well deterioration but also acts as a base from which the effi-

ciency of any remedial action can be measured. The common alternative to regular monitoring is to rely on the pump operator's memory and 'feel', but this 'seat of the pants' approach can be extremely misleading.

10.1 Incrustation and corrosion

The cause of incrustation is the change in physical and chemical conditions in the groundwater between the body of the aquifer and the well. The most common encrusting materials are car-

Fig. 10.1 Step drawdown test data. Shows the effects of well deterioration and later redevelopment.

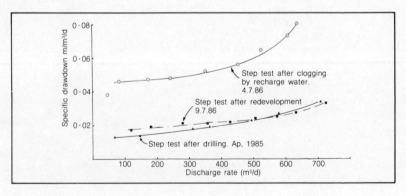

bonates and ferric hydroxide, but any incrustation is rarely composed of a single mineral. Calcium carbonate scale usually contains a proportion of magnesium carbonate, while the ferric hydroxide precipitate commonly contains some manganese and usually a fair percentage of carbonates.

Groundwater has sufficient residence time in most carbonate aquifers, or within porous aquifers cemented with carbonates, to become saturated with respect to bicarbonate ions: through solution of calcium (and magnesium) carbonate by weak carbonic acid. The carbonic acid is derived from atmospheric carbon dioxide dissolved in rain-water during recharge. The carbonate equilibrium equation:

$$Ca^{2+} + 2HCO_3^- = CO_2 + CaCO_3$$
$$\text{(soluble)} \qquad \text{(gas)} \; \text{(insoluble)}$$
$$+ \; H_2O \quad (1)$$

is held to the left, because the carbon dioxide gas cannot escape, owing to the pressure of confined groundwater. The confining pressure in the groundwater decreases as the water approaches the well, and this pressure gradient can be very high close to the well face. The sharp decrease in pressure may allow the exsolution of carbon dioxide, with resultant precipitation of calcium carbonate in the aquifer adjacent to the well as Eqn 1 moves to the right.

The phase-stability diagram for the solubility of iron minerals in water with respect to Eh and pH, is shown in Fig. 10.2. The natural conditions of most groundwaters are close to the

Fig. 10.2 Phase-stability diagram for solubility of iron.

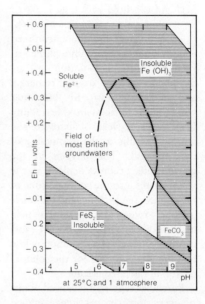

boundary between the fields of soluble (Fe^{2+}) and insoluble iron ($Fe(OH)_3$). Normally, the level of dissolved iron in groundwater is low—below 1mg/litre total iron, but slight changes in the water chemistry can increase the solubility of iron appreciably. A slight increase in acidity, through dissolved carbon dioxide or humic acids from peat, can cause an increase in iron content. When the water contains sufficient organic matter to cause the deoxygenation of the water, a low pH is accompanied by a negative Eh (redox potential) and the water may become high in iron, particularly if it is percolating through ferruginous sediments. This is seen in an extreme form with water polluted by organic-rich

129

leachates when the water becomes anoxic, has a negative Eh and a low pH, and can hold very high levels of iron; commonly several tens of mg/l.

Groundwater in the unconfined zone of aquifers normally will be oxygenated through solution of atmospheric oxygen during recharge, and will have a positive Eh. The water will lie in the ferric hydroxide stability field, so that the level of dissolved iron will be low due to the insolubility of that mineral. The oxygen in the groundwater becomes depleted as the water passes into the confined zone of the aquifer, the Eh falls, and the water gradually moves into the stability field of the soluble ferrous iron. In most confined aquifers, oxygenation is low and there is potential for the solution of iron in the ferrous state.

Boreholes represent localized points of access for oxygen into confined aquifers, and this local oxygenation, combined with an increase in pH through the exsolution of CO_2 can move the groundwater quality back into the stability field of ferric hydroxide. This change of stability, if there is sufficient iron dissolved in the water, can lead to the precipitation of ferric hydroxide where the oxygen meets native groundwater, that is, in the screen and in the formation immediately behind the screen.

The chemical precipitation of ferric hydroxide is most commonly accompanied by its biologically-mediated precipitation. Species of iron bacteria such as *Gallionella* derive energy from the oxidation of ferrous to ferric iron. The bacteria are filamentous, and occur in colonies in which the bacteria are coated in slimy sheaths. The solid ferric hydroxide resulting from the oxidation of soluble ferrous iron is precipitated as granules in the bacterial sheaths. The filamentous colonies and their ferruginous by-products can severely clog the screen and well face of a borehole.

The iron bacteria are very widespread in areas of ferruginous groundwaters, and it is difficult to avoid infection of a well. They can tolerate a range of oxygen levels in the water, but are most abundant where oxygen is freely available. It is not certain whether or not they live naturally in aquifers, but from their size—about 2 microns diameter—there is no doubt that they can invade to some depth into porous aquifers from infected wells.

The early precipitate of iron, both chemically and biologically precipitated, is soft and powdery. In biologically precipitated iron, carbon may be a significant percentage of the total iron incrustation. These deposits age with time, recrystallize and turn into a reddish brown, hard mixture of ferric oxides and hydroxides, which can cement a gravel pack like concrete.

The incrustation and cementation by both iron and carbonate minerals may entrap and incorporate fine materials moving out of the aquifer under the influence of pumping. This can make a bad situation worse.

A water well can also deteriorate through the contrasting—though chemically related—phenomenon of corrosion. Damage to the well structure can be from solution of the metal casing and screen, or from incrustation by the by-products of corrosion.

The whole problem of corrosion in groundwater is complex but, by and large, the main factors affecting corrosion are the nature of the metal corroded, and the physical and chemical condition of the water.

The most common process of corrosion is an electrochemical reaction, whereby anodic and cathodic areas are set up on the metal surface. Metal is dissolved from the anodic areas to yield positive ions in solution which, if the metal is steel and if the pH and Eh conditions are correct, will reprecipitate as ferric hydroxide (Fig. 10.3). This oxidation of the steel follows Eqn 2:

$$4\,Fe + 3O_2 + 6\,H_2O \rightarrow 4Fe(OH)_3 \quad (2)$$

The formation of electrolytic cells is encouraged by the presence of two different metals in contact. The process is enhanced if the metals are separated in the electromotive series (Table 10.1), which defines their relative susceptibility to corrosion, but it can operate where different grades of steel are in contact, for example at casing collars or at metallurgical discontinuities in the metal. The most important sites for corrosion in a water well are at physical imperfections on pipes, for example, the rough edges of screen slots or the screw threads of collars.

The main factors in water quality which affect corrosion are the pH and Eh (Fig. 10.2). The water in contact with steel has to be in the stability field of soluble Fe^{2+}, rather than the insoluble Fe(OH), for corrosion to be very active. Contributory, or related, factors include:

Salinity: High salinity encourages corrosion

Temperature: High temperature encourages corrosion

Oxygen: Anaerobic conditions cause low Eh and aid the growth of sulphate-reducing bacteria
Aerobic conditions provide the oxygen necessary for Eqn 2 to proceed

Carbon dioxide: High CO_2 levels lower the pH

Hydrogen sulphide: High H_2S levels indicate a low negative Eh

Organic acids: Acids from peat or pollution lower the pH and the oxygen level

The conditions suitable for corrosion can be very localized and lead to pitted corrosion, even in boreholes where the general conditions indicate a low corrosion potential. In an aerated borehole, iron dissolved by corrosion is precipitated very close to where it is dissolved (Eqn 2). In practice this appears as a blister of rust on the pipe surface. Beneath the blister, protected from the aerated well water, a micro-environment is set up where conditions may be extremely reducing and corrosion enhanced. Because the dissolved iron is continuously removed by

131

precipitation on the blister surface, corrosion also is continuous (Fig. 10.3).

Corrosion in anaerobic conditions may be accelerated through the actions of sulphate-reducing bacteria. These bacteria use the reduction of sulphate ions for their energy requirements. The totally anaerobic conditions, with very negative Eh required for the water to be in the stability field of FeS_2 are unusual in water wells, but may occur locally where water is polluted or in a normal well where pitted corrosion is taking place.

The danger from corrosion in a well is threefold; damage to the casing, damage to the screen and damage to the aquifer. Corrosion can perforate casing or cut through casing joints, so weakening the structure and allowing possibly polluted water into the well. The screen can corrode sufficiently to widen the slots and allow the gravel pack or formation to pass into the well. The products of corrosion, the precipitated ferric hydroxide, can clog both the screen and the formation in the same way as incrustation.

Table 10.1   Electromotive series of metals used in well construction

Increasing susceptibility to corrosion	↑	Magnesium Zinc Mild steel Brass bronze Stainless steels

Fig. 10.3   Pitted corrosion of steel casing (diagrammatic). After Campbell and Lehr (1973).

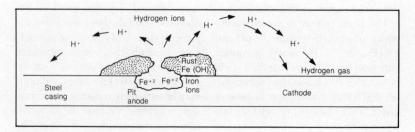

10.2 Maintenance measures

Across the world there are many thousands of production wells that have fallen into disuse through the breakdown of pumps and/or deterioration of the well structure through lack of maintenance. It is strongly recommended that priority is given to training technical staff in the monitoring of the state of a well and in the maintenance of both pumps and wells. This applies particularly to those countries in the arid or semi-arid areas of the world where there is a shortage of trained staff and where the breakdown of a water supply can be most serious. The steps in assessment of the status of a well and possible methods of restoration are summarized in Table 10.2.

The most effective measures against incrustation and corrosion in a well are those taken at the design stage. For example, a contributory factor to carbonate incrustation is the pressure-relief gradient as the water moves into a well; good screen design and well development to reduce this gradient will also reduce the likelihood of incrustation. The choice of a screen with a large open area (Section 3.3), to enable well maintenance to be undertaken effectively, is also important. Casing and screen can be protected from corrosion by cathodic protection, whereby sacrificial electrodes of a metal higher in the electromotive series than steel are provided; these electrodes are then corroded in preference to the casing. This system is used in some water wells to protect the pump installations, but is rarely used to protect the casing/screen string. Most commonly in water wells, corrosion-resistant materials are chosen at the design stage, if corrosive conditions are expected. These materials may be plastic, fibre-glass or resistant alloys (Chapter 3), or coated materials such as bitumen-coated casing and plastic-coated screen. The problem with coated materials is that if the coating is scratched during installation, as is likely, then the corrosion resistance is lost.

The maintenance of water wells, once installed, should be done on a regular basis of preventive maintenance. At present, this is rarely done, maintenance generally being in response to borehole failure. The problem with leaving a well until it fails completely is that by that time the screen and aquifer are totally blocked and any incrustation has had time to age. Development or well restoration depends on being able to get water into the clogged formation to remove the clogging material, and if blockage is total it may not be possible for water to enter the formation. Also if an incrustation has aged and recrystallized it will be much harder than in its early state and will be much more difficult to break up and remove.

Preventive maintenance can be undertaken on an arbitrary time scale, or when signs of deterioration *first* appear. In a low-risk situation a maintenance work-over may be done every few years—say five years, while in an area where incrustation is prevalent, then an annual work-over may be required. It is recommended here that the performance of a well is monitored annually by a step-

133

Table 10.2 Assessment of the need for well restoration

---

1. The pump discharge stops: then there is likely to be a pump failure.

    Action: Remove and service or replace the pump.

2. The pumping water level is constant but the discharge rate falls: Then the pump is likely to be damaged or to becoming progressively clogged.

    Action: Remove and service or replace the pump. Examine cause of damage or clogging and try to remedy. For example, damage to the pump through sand pumping may require the well to be screened or re-screened (Chapter 3). Run a geophysical survey with CCTV to assess the condition of the screen or well face.

3. The pumping water level has fallen (shown by direct measurement or because the pump begins pumping air).

    Action: Check by comparison with original pumping-test data and later water-level monitoring (Chapter 9) if the fall in water level has been (a) continuous or (b) sudden or progressively greater.

    If (a), then a continuously falling water level suggests that the aquifer is being overdrawn rather than the well being damaged. The only remedy is to reduce the discharge from the whole aquifer.

    If (b), then a sudden or increasing rate of drawdown suggests that the well is being clogged by incrustation of the well face or screen.

    Action: Conduct a step drawdown test and compare the results with the original test when the well was drilled. If the performance is reduced, then re-develop the well as in Table 8.2.

4. Well appears satisfactory.

    Action: Run step drawdown test (Section 9.1.1) to compare with the original test and to confirm the well is satisfactory, or to provide data against which future performance can be measured. Do this regularly (p.103), at least every five years.

---

drawdown test, and that a well work-over is done when the tests show that the well performance is dropping off.

The methods used for well restoration after incrustation are the same as those for well development, except that in well restoration the incrustation has to be broken up before it can be removed. Carbonate incrustation can be removed by acidization, as with the development of limestone water wells but, if the incrustation is heavy, the restoration will be improved by initial very high-pressure jetting to break up the carbonate and allow the acid better access. Acidization can also dissolve ferric hydroxide precipitate in its early powdery form, but the solubility of this mineral is much less when it has recrystallized and hardened. With hardened incrustation of iron hydroxide, restoration has to depend on physically breaking up the cemented formation by very high-pressure jetting or similarly energetic methods. Care has to be taken when using vigorous methods, not to damage the well and make the situation worse. Very high-pressure jetting, for example, can cut through plastic casing and can only be used safely in steel casing. Shocking a well by explosives similarly can disrupt weak casing or screen.

The restoration of a well should comprise three stages: the break up of the incrustation, the removal of the detritus by surging and clearance pumping, and finally a step-drawdown test. The latter indicates whether the restoration has been successful, and acts as the base against which future well performance can be measured. With a well in which iron incrustation has taken place, the well should be sterilized between the clearance pumping and the step test. Sterilization ensures that the ubiquitous iron bacteria are destroyed and delays re-infestation of the well.

Well restoration or maintenance can produce quantities of sludge or severely contaminated water. Measures must be taken for temporary storage of this material on-site, and its ultimate disposal to a waste disposal site or sewage works.

A water well damaged by corrosion to such an extent that the casing is perforated or that it is pumping sand, can only be restored by total or partial re-lining. The necessary course of action may be decided after a comprehensive geophysical logging programme (Section 7.3). Holes in the casing can be revealed by features on the differential temperature and conductivity logs, by flow logs or by resistivity and casing collar logs. The nature and extent of the perforations can be shown by closed-circuit television inspection, if incrustation is light and the water column is clear.

The casing used for re-lining should be chosen to avoid a repetition of the problem, that is, it should be corrosion resistant. Seals between the new casing and the corroded section must be sound to avoid seepage of polluting water between the casings. The seals could be plastic rings or malleable metal swedge rings. If the casing being re-lined is the pump chamber of a well, then the size of pump which can be installed will be reduced. Corroded screen should not be re-lined if possible, because concentric screens can in-

135

duce turbulence and physical abrasion in the annulus between. The corroded screen should be removed, if possible, and replaced with corrosion-resistant screen.

References

1 Books for the small field library

ANDERSON, K. E. (Ed) (1973) *Water well handbook*. Rollo Missouri, Missouri Water Well and Pump Contractors Assocn, Inc.

DHV (1979) *Shallow Wells*. Amersfoort, The Netherlands, DHV Consulting Engineers.

DRISCOLL, F. G. (1986) *Groundwater and Wells* (2nd Ed.), St Paul, Minnesota, Johnson Division.

FREEZE, R. A. & CHERRY, J. A. (1979) *Groundwater*. New Jersey, Prentice-Hall.

HEM, J. D. (1985) *Study and Interpretation of the Chemical Characteristics of Natural Water*. US Geological Survey Water-Supply Paper 2254 (3rd Ed). Washington, US Government Printing Office.

KRUSEMAN, G. P. & DE RIDDER, N. A. (1970) *Analysis and Evaluation of Pumping Test Data*. Wageningen, Netherlands, Bulletin 11. International Institute for Land Reclamation and Improvement.

2 Books for design specifications

AMERICAN PETROLEUM INSTITUTE (1984) *API Specification for Casing, Tubing, and Drill Pipe*. API Spec. 5A Thirty-Seventh Ed. Dallas, API.

BRITISH STANDARDS INSTITUTION (1965) *Specification for Water Well Casing*. BS 879: 1985, London, BSI.

BRITISH STANDARDS INSTITUTION (1974) *Core Drilling Equipment. Part 1. Basic Equipment*. BS 4019. Part 1. 1974, London, BSI.

BRITISH STANDARDS INSTITUTION (1985) *Methods of Testing for Soil for Civil Engineering Purposes*. BS 1377: 1975 Test 7a, London, BSI.

BRITISH STANDARDS INSTITUTION (1981) *Methods of Measurement of Liquid Flow in Open Channels. Part 4. Weirs and Flumes. Part 4A. Thin-plate Weirs*. BS 3680: Part 4A: 1981, London, BSI.

BRITISH STANDARDS INSTITUTION (1981) *Code of Practice for Site Investigations*. BS 5930: 1981, London, BSI.

BRITISH STANDARDS INSTITUTION (1983) *Code of Practice for Test Pumping Water Wells*. BS 6316: 1983, London, BSI.

137

3 Publications for general background reading

AHMAD, N. (1979) *Tubewell Theory and Practice*. Lahore, Pakistan, Shahzad Nazir.

BAKIEWICZ, W., MILNE, D. M. & PATTLE, A. D. (1985) 'Development of public tubewell designs in Pakistan', *Q. J. Engin. Geol.* 18, 63–77.

BOULTON, N. S. (1963) 'Analysis of data from non-equilibrium pumping tests allowing for delayed yield from storage ', *Proc. Inst. Civil Enginrs*, 26, 469–482.

BOUWER, H. (1978) *Groundwater Hydrology*. Tokyo, McGraw-Hill Kogakusha.

BRASSINGTON, R (1988) *Field Hydrogeology*. Open University Press.

CAMPBELL, M. D. & LEHR, J. H. (1973) *Water Well Technology*. New York, McGraw-Hill.

CLARK, L. (1977) 'The analysis and planning of step drawdown tests', *Q. J. Engin. Geol.* 10, 125–143.

CLARK, L. (1985) 'Groundwater abstraction from Basement Complex areas of Africa', *Q. J. Engin. Geol.* 18, 25–34.

CLARK, L. & STONER, R. F. (1980) 'Regional groundwater development in temperate and arid zones', *Proc. of Conference: Water resources, a changing strategy.'* London, Institute of Civil Engineers.

CRUSE, K. (1979) 'A review of water well drilling methods', *Q. J. Engin. Geol.* 12, 79–95.

DAVIS, S. N. & De Weist, R. J. M. (1966) *Hydrogeology*. New York, J. Wiley.

DUMBLETON, J. E. (1953) *Wells and Boreholes for Water Supply*. London, The Technical Press.

GIBSON, U. P. & SINGER, R. D. (1971) *Water Well Manual*. Berkeley, Californias, Premier Press.

HAMILL, L. & BELL, F. G. (1986) *Groundwater Resource Development*. Sevenoaks, Kent, Butterworths.

HITCHMAN, S. P. (1983) A guide to the field analysis of groundwater: report FLPU 83–12. Institute of Geological Sciences. Keyworth. England.

HUNTER BLAIR, A. (1968) *Well Screens and Gravel Packs*. Technical Paper TP 64. Medmenham, Bucks, UK, Water Research Association.

JACOB, C. E. (1946) 'Drawdown test to determine effective radius of artesian well', *Trans. Amer. Soc. Civil Engnrs* 112, 1047–1070.

JOHNSON, E. E., Inc. (1966) *Groundwater and Wells*. St Paul, Minnesota, Edward E. Johnson Inc. (See also Driscoll, 1986).

NAYLOR, J. A., ROWLAND, C. D., YOUNG, C. P. & BARBER, C. (1978) *The Investigation of Landfill Sites*. Technical Report TR 91. Medmenham, Bucks, Water Research Centre.

NOLD, J. F. (1980) *The Nold Well Screen Book*. Ed. Bieske, E. & Wandt, K Stockstadt am Rhein, J. F. Nold & Co.

PIRSON, S. J. (1963) *Handbook of Well Log Analysis*. Englewood Cliffs, Prentice-Hall.

RUSHTON, K. R. & REDSHAW, S. C. (1979) *Numerical Analysis by Analog and Digital methods: Seepage and Groundwater Flow*. Chichester, J. Wiley.

STONER, R. F., MILNE, D. M. & LUND, P. J. (1979) 'Economic design of wells', *Q. J. Engin. Geol.* 12, 63–78.

TERZAGHI, K. & PECK, R. B. (1948) (1967, 2nd En) *Soil Mechanics in Engineering Practice*. New York, J. Wiley.

References

THEIS, C. V. (1935) 'The relation between the lowering of the piezometric surface and the rate and duration of discharge of a well using ground-water storage', *Trans. Am. Geophys. Union*, 16, 519–524.

THIEM, G. (1906) *Hydrologische Methoden*. Leipzig, East Germany, Gebhart.

TODD, D. K. (1980) *Groundwater Hydrology*. J. Wiley.

TUCKER, M. E. (1982) *The Field Description of Sedimentary Rocks*. Geological Society of London Handbook Series, Milton Keynes, Open University Press.

WALTON, W. C. (1962) *Selected Analytical Methods for Well and Aquifer Evaluation*. Illinois State Water Survey, Bulletin No. 49. Urbana, Illinois.

WALTON, W. C. (1970) *Groundwater Resource Evaluation*. New York, McGraw-Hill.

WATT, S. R. & WOOD, W. E. (1977) *Hand Dug Wells and their Construction*. London, Intermediate Technology Publications Ltd.

WILLIAMS, B. B. GIDLEY, J. L. & SCHECHTER, R. S. (1979) *Acidizing Fundamentals*. Monograph Volume 6, Henry L. Doherty Series. 124 pp. New York & Dallas, Society of Petroleum Engineers of the American Institute of Mining, Metallurgical and Petroleum Engineers, Inc.

Appendices

Appendix A.I. Units

Water wells and boreholes, their design, construction and testing should be specified and described in SI units (Système International d'Unités) wherever possible. The SI system of units was agreed in 1960, and is based on the metre (m), kilogram (kg), second (s), Ampere (A), Kelvin (K), mole (mol) and candela (cd). These units have been adopted generally by the scientific community, but in the field of boreholes and wells there are also three competing systems of traditional units: Imperial, American, and metric. In any given country, one of these systems, or a mixture of the units, may have been adopted. Tables to allow conversion from any system to SI units are given below.

There are problems in conversion to SI units in specific areas where the use of traditional units is particularly entrenched.

Volumes of liquids. The American and Imperial systems both use the word 'gallon' but the US gallon (US gal.) is only 0.8326 Imperial gallons (Imp. gal.). In any volumetric conver-

sion, care must be taken to identify which kind of gallon is being dealt with. In the metric system the unit of volume is the litre, and the litre (l or L) was accepted into the SI system in 1964 as 1 litre = 1 cubic decimetre $(1 \cdot 0 \times 10^{-3} \, \text{m}^3)$.

Length, diameter and mass are specified in older British Standard Institution (BSI) specifications in feet (ft), inches (in) and pounds respectively. Oil-well casing is specified in similar units by the American Petroleum Institute (API) 1984.

Aquifer characteristics. The SI unit of aquifer transmissivity is m^2/s, but m^2/d is more commonly used in hydrogeology. Similarly, m/d is used for hydraulic conductivity rather than m/s. This use of day does generally yield values of the aquifer characteristics in conveniently sized whole numbers.

Discharge rates. The unit m^3/s (cumec) is a convenient unit for surface flows but is too large to be convenient for borehole discharge rates. The unit m^3/d, although not truly an SI unit, conforms with the m^2/d and m/d of transmissivity and hydraulic conductivity.

A.I.1   Conversion tables

Length (SI unit, metre, m)

	m	ft	in
1 m	1·000	3·281	39·37
1 ft	0·3048	1·000	12·00
1 in	$2·540 \times 10^{-2}$	$8·333 \times 10^{-2}$	1·000

Area (SI unit, square metre, m^2)

	m^2	ft^2	$acre$	$hectare$
1 m²	1·000	10·76	$2·471 \times 10^{-4}$	$1·0 \times 10^{-4}$
1 ft²	$9·29 \times 10^{-2}$	1·000	$2·29 \times 10^{-5}$	$9·29 \times 10^{-5}$
1 acre	$4·047 \times 10^{2}$	$4·356 \times 10^{4}$	1·000	$4·047 \times 10^{-1}$
1 hectare	$1·0 \times 10^{4}$	$1·076 \times 10^{5}$	2·471	1·000

Volume (SI unit, cubic metre, m^3)

	m^3	l	$Imp. gal$	$US gal$	ft^3
1 m³	1·000	$1·000 \times 10^{3}$	$2·200 \times 10^{2}$	$2·642 \times 10^{2}$	35·32
1 l	$1·000 \times 10^{-3}$	1·000	0·2200	0·2642	$3·532 \times 10^{-2}$
1 Imp. gal	$4·546 \times 10^{-3}$	4·546	1·000	1·200	0·1605
1 US gal	$3·785 \times 10^{-3}$	3·785	0·8326	1·000	0·1337
1 ft³	$2·827 \times 10^{-2}$	28·27	6·229	7·480	1·000

Time   (SI unit, second, s)

	s	min	h	d
1 s	1·000	1·667×10⁻²	2·777×10⁻⁴	1·57×10⁻⁵
1 min	60·00	1·000	1·667×10⁻²	6·944×10⁻⁴
1 h	3·600×10³	60·00	1·000	4·167×10⁻²
1 d	8·640×10⁴	1·440×10³	24·00	1·000

Discharge rate   (SI unit, cubic metre per second, m³/s)

	m³/s	m³/d	l/s	Imp. gal/d	US gal/d	ft³/s
1 m³/s	1·000	8·640×10⁴	1·000×10³	1·901×10⁷	2·282×10⁷	35·313
1 m³/d	1·157×10⁻⁵	1·000	1·157×10⁻²	2·200×10²	2·642×10²	4·087×10⁻⁴
1 l/s	1·000×10⁻³	86·40	1·000	1·901×10⁴	2·282×10⁴	3·531×10⁻²
1 Imp.gal/d	5·262×10⁻⁸	4·546×10⁻³	5·262×10⁻⁵	1·000	1·201	1·858×10⁻⁶
1 US gal/d	4·381×10⁻⁸	3·785×10⁻³	4·381×10⁻³	0·8327	1·000	1·547×10⁻⁶
1 ft³/s	2·832×10⁻²	2·447×10³	2·832×10⁴	5·382×10⁵	6·464×10⁵	1·000

Hydraulic conductivity   (SI unit, cubic metre per second per square metre, m³/s/m² or m/s)

	m/s	m/d	Imp. gal/d-ft²	US gal/d-ft²	ft/s
1 m/s	1·000	8·640×10⁴	1·766×10⁶	2·12×10⁶	3·80×10²
1 m/d	1·157×10⁻⁵	1·000	20·44	24·54	4·398×10⁻³
1 Imp gal/d-ft²	5·663×10⁻⁷	4·893×10⁻²	1·000	1·201	1·858×10⁻⁶
1 US gal/d-ft²	4·716×10⁻⁷	4·075×10⁻²	0·8327	1·000	1·547×10⁻⁶
1 ft/s	2·632×10⁻³	2·273×10²	5·382×10⁵	6·463×10⁵	1·000

Transmissivity (SI unit, cubic metre per second per metre, $m^3/s/m$ or m^3/s)

	m^2/s	m^2/d	Imp. gal/d-ft	US gal/d-ft	ft^2/s
1 m^2/s	1·000	$8·640 \times 10^4$	$5·793 \times 10^6$	$6·957 \times 10^6$	$1·161 \times 10^2$
1 m^2/d	$1·157 \times 10^{-5}$	1·000	67·05	80·52	$1·343 \times 10^{-3}$
1 Imp.gal/d-ft	$1·726 \times 10^{-7}$	$1·491 \times 10^{-2}$	1·000	1·201	$1·858 \times 10^{-6}$
1 US gal/d-ft	$1·437 \times 10^{-7}$	$1·242 \times 10^{-2}$	0·8326	1·000	$1·547 \times 10^{-6}$
1 ft^2/s	$8·617 \times 10^{-3}$	$7·445 \times 10^2$	$5·382 \times 10^5$	$6·463 \times 10^5$	1·000

Mass (SI unit, kilogram, kg)
1 kg = 2·205 pounds mass.

Force (SI unit, newton, N)
1 N = 0·2248 pounds force.

Pressure and stress (SI unit, pascal, Pa = 1 Nm^{-2})
1 megapascal (MPa) = 145 pounds force per square inch (psi).
101,325 pascals = 1 standard atmosphere (atm) = 1·01325 bar.
9806·65 pascals = 1 m head of water = 10·33 atm.
2989 pascals = 1 ft head of water.

143

Appendix A.II. Rotary drilling equipment

A.II.1 Standard core barrel sizes

Size symbol	Nominal hole diameter (mm)	Core diameter G, M & F series (mm)	T series (mm)	Standard core barrel designs WF	WG	WM	WT
R	30		19	—	—	—	RWT
E	38	21	23	EWF	EWG	EWM	EWT
A	48	30	33	AWF	AWG	AWM	AWT
B	60	42	44	BWF	BWG	BWM	BWT
N	76	55	59	NWF	NWG	NWM	NWT
H	99	76	81	HWF	HWG	—	HWT
P	120	92		PWF			
S	146	113		SWF			
U	175	140		UWF			
Z	200	165		ZWF			

The first letter of the core barrel design is the size symbol. The second letter, W, signifies that the core barrel head is threaded to fit W design drill rods. F, G, M and T are design series of the core barrels:

F is a face discharge core barrel:
G is a general design.
M is a development of the G series for coring friable rocks.
T is a thin-walled barrel and cuts a slightly larger core.

The size symbols ensure compatibility of parts: a B size drill rod will fit a B size core barrel, which will pass through N size casing and drill a hole capable of receiving A size casing.

A.II.1   Drilling fluid information

Bentonite mix for drilling mud:

Per cent mix bentonite:water	Approximate number of 100 lb (45 kg) sacks of bentonite to make volume of mud			
	50 m³	100 m³	150 m³	200 m³
4	44	88	133	176
5	56	112	167	224
6	67	134	200	268
7	78	156	233	311

Normal mud is a 5–6% mix, but a stronger mix may be needed with caving formations. For assistance in percussion drilling an approximately 6% mix is recommended (about 1.25 bags to 1 m³ of water).

Velocity of mud return up borehole:

Drill pipe diameter		Nominal borehole diameter		Mud circulation rate			
				15 l/s	30 l/s	45 l/s	60 l/s
mm	in	mm	in				
89	3½	150	6	1·33	2·65	3·98	5·31
		200	8	0·60	1·20	1·80	2·39
		250	10	0·35	0·70	1·05	1·40
114	4½	200	8	0·71	1·41	2·12	2·83
		250	10	0·39	0·77	1·16	1·54
		300	12	0·25	0·50	0·74	0·99
141	5⁹⁄₁₆	250	10	0·63	1·27	1·90	2·53
		300	12	0·27	0·54	0·81	1·08
		350	14	0·19	0·37	0·56	0·74
168	6⅝	300	12	0·31	0·62	0·93	1·24
		350	14	0·23	0·45	0·68	0·91
		400	16	0·14	0·29	0·43	0·58
		450	18	0·11	0·22	0·33	0·44
		500	20	0·09	0·17	0·26	0·34

Mud velocity is in m/s.

Polymer drilling muds.   The properties of polymer muds depend on the manufacturer's formulation, therefore the following figures should be used as a guide only.

Concentration	Normal rotary drilling	— *c.* 1·0 kg per m³ of mud
	Caving formation	— 1·0–2·0 kg per m of mud
	Clay formation	— Up to 0·5 kg per m³ of mud
Breakdown	Rapid. Hypochlorite solution (12% Cl) at about 0·4% mix	
	Slow. Hydrogen peroxide (110 vol.) at about 0·4% mix.	
Extension	The viscosity of polymer drilling mud can be maintained beyond its normal life by food-grade inhibitors.	

The manufacturer's guidelines to the use of polymer muds, particularly with respect to well disinfection, must be followed closely.

Foam additives. The properties of foaming agents depend on the manufacturer's formulation. For guidance, a mix of about 0.4% should make a thick workable foam. A particularly viscous foam can be made using a mix of around 5% foaming agent and 1% polymer drilling mud.

Air (see Fig. A.II.1) The volume of compressed air required for air-flush drilling depends on the borehole diameter, the hole depth and the diameter of the drill stem being used. Figure A.II.1 gives guidance on the volume of air required in the conditions most commonly met while drilling exploration boreholes or water wells.

Fig. A.II.1 Air requirements for air-flush drilling. After Anderson (1973).

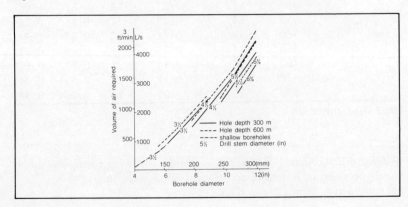

Appendix A.III Water-quality conversion table

Determinand	Milli-equivalent to milligram per litre	Milligram per litre to milli-equivalent
Calcium	20·04	0·04991
Magnesium	12·16	0·08224
Potassium	39·10	0·02558
Sodium	23·00	0·04348
Bicarbonate	61·01	0·01639
Sulphate	48·04	0·02082
Chloride	35·46	0·02820
Nitrate	62·01	0·01613

Nitrate as mg nitrogen/l = Nitrate as mg nitrate/l × 0·2258.
Bicarbonate as mg/l = alkalinity in mg/l as $CaCO_3$ × 1·22.

Appendix A.IV Discharge measurements

A.IV.1 *Discharge over a 90° V-notch weir*

Head (mm)	Discharge (m^3/d)	Head (mm)	Discharge (m^3/d)	Head (mm)	Discharge (m^3/d)
30	20	130	731	230	3024
40	37	135	803	240	3367
50	66	140	878	250	3729
60	109	145	958	260	4113
65	132	150	1043	270	4520
70	159	155	1131	280	4951
75	188	160	1224	290	5405
80	220	165	1322	300	5884
85	256	170	1424	310	6388
90	295	175	1530	320	6917
95	337	180	1642	330	7470
100	382	185	1758	340	8050
105	431	190	1879	350	8657
110	483	195	2005	360	9290
115	539	200	2134	370	9950
120	599	210	2411	380	10,638
125	663	220	2704		

Discharge tables for V-Notch Weirs with other angles are given in BS 3680: Part 4A: 1981 or Anderson, 1973.

A.IV.2 Discharge tables for orifice plates

Head (mm)	75 mm orifice		100 mm orifice		125 mm orifice	
	100 mm pipe	150 mm pipe	150 mm pipe	200 mm pipe	150 mm pipe	200 mm pipe
100	503					
200	657	505	999	916	1897	1515
300	786	615	1214	1123	2333	1869
400	901	710	1387	1283	2681	2191
500	1007	798	1538	1436	2954	2414
600	1109	883	1677	1572	3216	2632
700	1199	955	1807	1692	3499	2829
800	1289	1021	1922	1807	3728	2932
900	1367	1078	2052	1910	3995	3150
1000	1435	1133	2162	2004	4218	3325
1100	1499	1182	2264	2095	4425	3504
1200	1565	1236	2370	2185	4622	3646
1300	1630	1284	2467	2275	4796	3782
1400	1688	1338	2554	2357	4976	3940
1500	1739	1380	2639	2445	5150	4066
1600	1794	1435	2718	2528	5276	4207
1700	1852	1485	2802	2609	5439	4316

Head (mm)	150 mm orifice		175 mm orifice	200 mm orifice
	200 mm pipe	*250 mm pipe*	*250 mm pipe*	*250 mm pipe*
200	2463	2126	3226	5069
300	2998	2649	3946	6066
400	3434	3057	4534	6878
500	3831	3466	5047	7630
600	4180	3728	5532	8284
700	4513	4055	5979	8884
800	4796	4316	6415	9521
900	5052	4578	6774	10,083
1000	5276	4785	7096	
1100	5564	5003	7412	
1200	5777	5205	7739	
1300	6017	5412	8066	
1400	6224	5603	8371	
1500	6442	5777	8655	
1600	6649	5968		
1700	6840	6164		

Discharge units: m³/d.

Appendix A.V Sample descriptions

A.V.1 Sieve analyses

The grain-size distribution of a sediment is measured by sieve analyses (BS 1377: 1975: Test 7a). Samples of about 500 g should be taken from undisturbed core if possible or, if not, then from the least-contaminated disturbed sample from the target aquifer. Each sample must be dried, preferably in an oven, and then weighed to the nearest gram. Impure sands may dry into clay-bound crumbs, in which case the sample should be disassociated by gentle grinding in a mortar. The sample is then passed through a nest of sieves of decreasing

Fig. A.V.1 Mechanical sieve shaker.

mesh size, and shaken until each sieve only holds a clean subsample. The sieves may be shaken on a mechanical shaker (Fig. A.V.1) or by hand. In very clayey samples dry sieving may not break up the crumbs, and each sieve in turn will then need to be washed to get a clean fraction. The nest of sieves should be chosen to give a spread of grain sizes; a minimum nest comprising the following sizes is recommended for aquifer material: 2.0 mm, 1.0 mm, 0.5 mm, 0.25 mm, 0.125 mm and 0.63 mm. The fraction retained in each sieve is weighed, converted into a per cent of the total sample weight, and then the cumulative results are plotted as a grain-size distribution graph (Fig. A.V.2). As a check on the analysis, the sum of the weight of the fractions should equal the weight of the original sample.

Sieve analyses are best done in a laboratory, but they can be done in the field with limited facilities. The samples then will have to be sun-dried and sieved by hand shaking, and the sand fractions will have to be weighed using a mechanical balance.

The cumulative weight retained is plotted against the sieve mesh size in mm or in phi units, as on Fig. A.V.2. The grain size on the curve is commonly referred to as the percentile passing; for example, the D60 size of a sample is the sieve size through which 60% of the sample will pass. On Fig. A.V.2. this is 0.09 mm. The sorting coefficient of a sediment is the D60 divided by the D10 size, and a well-sorted sediment will have a sorting coefficient of less than 2.5, while a badly sorted sediment will have a coefficient of over 5.

Fig. A.V.2 Grain-size distribution curve.

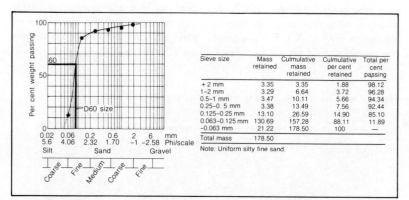

Sieve size	Mass retained	Culmulative mass retained	Cumulative per cent retained	Total per cent passing
+2 mm	3.35	3.35	1.88	98.12
1–2 mm	3.29	6.64	3.72	96.28
0.5–1 mm	3.47	10.11	5.66	94.34
0.25–0.5 mm	3.38	13.49	7.56	92.44
0.125–0.25 mm	13.10	26.59	14.90	85.10
0.063–0.125 mm	130.69	157.28	88.11	11.89
–0.063 mm	21.22	178.50	100	—
Total mass	178.50			

Note: Uniform silty fine sand.

The design criteria for gravel packs and screens are most commonly defined in terms of the D sizes of the aquifer and pack materials (Figs 3.3 and 3.4).

A.V.2 Sample description aids

In lithological logging the grains of a sediment are described in terms of their roundness and sphericity. Charts of different degrees of roundness and sphericity, to help the estimation of these characteristics, are given in Fig. A.V.3. Grains to be described should be examined by lens or low-powered microscope and their shape compared with those on the charts.

Sediments are commonly classified by their sorting, and charts for the estimation of sorting are given on Fig. A.V.4. Disturbed formation samples from drilling returns are made up of different proportions of various rock types, and one has to estimate the percentages of these rock types to build

Fig. A.V.3 Charts for estimating grain roundness and sphericity. From Tucker (1982).

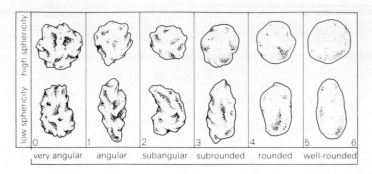

151

up a percentage log (Fig. 7.3). Charts to help estimate percentages in samples are shown on Fig. A.V.5. Enough of the sample should be put in a Petrie dish, so that the bottom is barely covered, and then the abundance of various grain types compared visually with the charts.

Fig. A.V.4 Charts for estimating sediment sorting. From Tucker (1982).

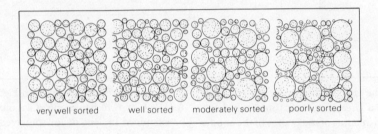

Fig. A.V.5 Charts for estimating percentages in disturbed formation samples.

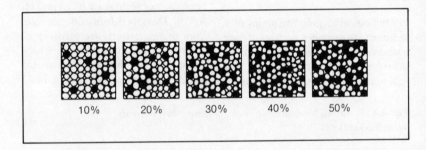

Index